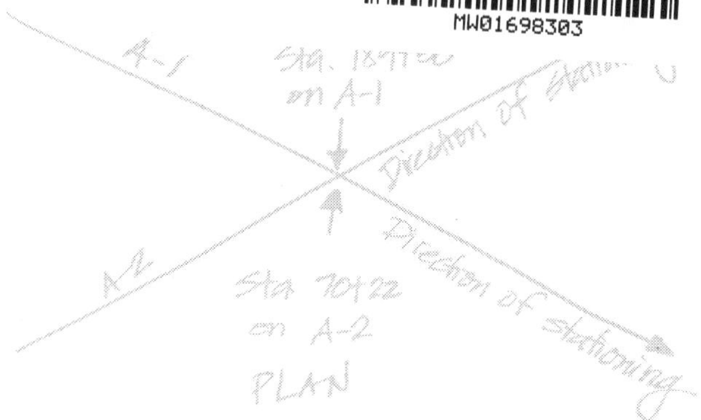

PE

principles and practice
of engineering

$y = \frac{A}{200L}x^2 \quad x = 7{,}200 - 7{,}022 = 178 \text{ ft}$

$y = \frac{(-2-8)}{200(1{,}000)}(178)^2 = 0.79$

Elevation on Curve at Station 70+22 = 112.66 + 0.79 = 113.45 ft

Alt. Solution:

$Sta_{PVC} = (67+00) - \frac{L}{2} = (67+00) - \frac{1000}{2} = (62+00)$

$x = (70+22) - (62+00) = 822 \text{ ft}$

$Y_{PVC} = 103.00 + 0.02\left(\frac{1{,}000}{2}\right) = 113.0$

$Y_x = Y_{PVC} + g_1 x + \left(\frac{g_2 - g_1}{2L}\right)x^2$

$= 113.00 + (-0.02)(822) + \left(\frac{0.08 - (-0.02)}{2{,}000}\right)822^2$

$= 113.45 \text{ ft}$

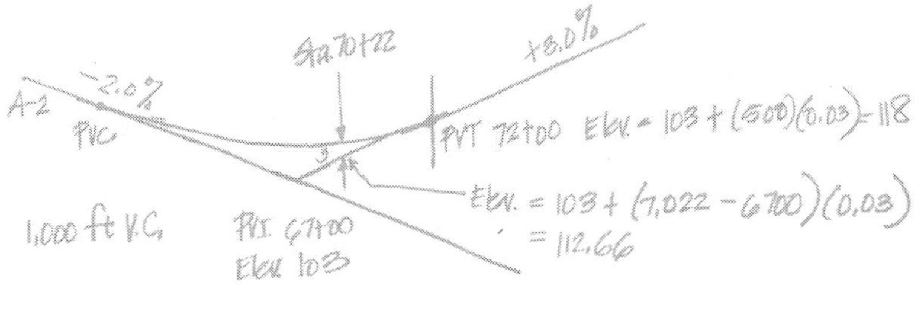

NCEES
advancing licensure for engineers and surveyors

civil: TRANSPORTATION

sample questions + solutions

P.O. Box 1686
Clemson, SC 29633
800-250-3196
www.ncees.org

ISBN 978-1-932613-64-3

Printed in the United States of America
November 2010

Paper used in the printing of NCEES study materials meets environmental guidelines of the Rainforest Alliance, Elemental Chlorine Free (ECF) papers, and Sustainable Forestry Initiative (SFI).

CONTENTS

About the publisher

NCEES is a national nonprofit organization dedicated to advancing professional licensure for engineers and surveyors. It develops, administers, and scores the examinations used for engineering and surveying licensure in the United States. It also facilitates professional mobility and promotes uniformity of the U.S. licensure processes through services for its member licensing boards and licensees. These services include the records program, study materials, credentials evaluations, exam administration, and more.

NCEES is composed of member licensing boards from throughout the United States and its territories.

Updates on exam content and procedures

NCEES.org is our home on the Web. Visit us there for updates on everything exam-related, including specifications, exam-day policies, scoring, and corrections to published study materials. This is also where you will register for the exam and find the additional steps you should follow in your state to be approved for the exam.

Exam-day schedule

Be sure to arrive at the exam site on time. Late-arriving examinees will not be allowed into the exam room once the proctor has begun to read the exam script. The report time for the exam will be printed on your Exam Authorization. Normally, you will be given 1 hour between morning and afternoon sessions.

Admission to the exam site

To be admitted to the exam, you must bring two items: (1) your Exam Authorization and (2) a current, signed, government-issued photo identification.

Candidate Agreement

The NCEES Candidate Agreement describes references and other personal materials allowed in the exam room. It also describes other exam policies designed to protect exam content. This document will be printed in your exam book. It is also available for download at NCEES.org. If you register to take an NCEES exam, you are strongly encouraged to review this document prior to exam day, as you will be asked to affirm that you abide by the policies and procedures it describes.

Scoring and reporting

NCEES typically releases exam results to its member licensing boards 8–10 weeks after the exam. Depending on your state, you will be notified of your exam result online through your My NCEES account or via postal mail from your state licensing board. Detailed information on the scoring process can be found at NCEES.org.

EXAM SPECIFICATIONS

NCEES Principles and Practice of Engineering
CIVIL BREADTH and TRANSPORATION DEPTH
Exam Specifications

Effective Beginning with the April 2008 Examinations

- The civil exam is a breadth and depth examination. This means that examinees work the breadth (AM) exam and one of the five depth (PM) exams.
- The five areas covered in the civil exam are construction, geotechnical, structural, transportation, and water resources and environmental. The breadth exam contains questions from all five areas of civil engineering. The depth exams focus more closely on a single area of practice in civil engineering.
- Examinees work all questions in the morning session and all questions in the afternoon module they have chosen. Depth results are combined with breadth results for final score.
- The exam is an 8-hour open-book exam. It contains 40 multiple-choice questions in the 4-hour AM session, and 40 multiple-choice questions in the 4-hour PM session.
- The exam uses both the International System of Units (SI) and the US Customary System (USCS).
- The exam is developed with questions that will require a variety of approaches and methodologies, including design, analysis, and application. Some problems may require knowledge of engineering economics.
- The knowledge areas specified as examples of kinds of knowledge are not exclusive or exhaustive categories.
- The specifications for the **AM exam** and the **Transportation PM exam** are included here. The **design standards** applicable to the Transportation PM exam are shown on **ncees.org**.

CIVIL BREADTH Exam Specifications

	Approximate Percentage of AM Exam
I. Construction	**20%**

A. Earthwork Construction and Layout
   1. Excavation and embankment (cut and fill)
   2. Borrow pit volumes
   3. Site layout and control

B. Estimating Quantities and Costs
   1. Quantity take-off methods
   2. Cost estimating

C. Scheduling
   1. Construction sequencing
   2. Resource scheduling
   3. Time-cost trade-off

D. Material Quality Control and Production
   1. Material testing (e.g., concrete, soil, asphalt)

E. Temporary Structures
   1. Construction loads

II. Geotechnical **20%**

A. Subsurface Exploration and Sampling
   1. Soil classification
   2. Boring log interpretation (e.g., soil profile)

B. Engineering Properties of Soils and Materials
   1. Permeability
   2. Pavement design criteria

C. Soil Mechanics Analysis
   1. Pressure distribution
   2. Lateral earth pressure
   3. Consolidation
   4. Compaction
   5. Effective and total stresses

D. Earth Structures
   1. Slope stability
   2. Slabs-on-grade

E. Shallow Foundations
   1. Bearing capacity
   2. Settlement

F. Earth Retaining Structures
   1. Gravity walls
   2. Cantilever walls
   3. Stability analysis
   4. Braced and anchored excavations

III. Structural **20%**

A. Loadings
   1. Dead loads
   2. Live loads
   3. Construction loads

B. Analysis
   1. Determinate analysis

C. Mechanics of Materials
   1. Shear diagrams
   2. Moment diagrams
   3. Flexure
   4. Shear
   5. Tension
   6. Compression
   7. Combined stresses
   8. Deflection

D. Materials
   1. Concrete (plain, reinforced)
   2. Structural steel (structural, light gage, reinforcing)

E. Member Design
   1. Beams
   2. Slabs
   3. Footings

IV. Transportation **20%**

A. Geometric Design
1. Horizontal curves
2. Vertical curves
3. Sight distance
4. Superelevation
5. Vertical and/or horizontal clearances
6. Acceleration and deceleration

V. Water Resources and Environmental **20%**

A. Hydraulics – Closed Conduit
1. Energy and/or continuity equation (e.g., Bernoulli)
2. Pressure conduit (e.g., single pipe, force mains)
3. Closed pipe flow equations including Hazen-Williams, Darcy-Weisbach Equation
4. Friction and/or minor losses
5. Pipe network analysis (e.g., pipeline design, branch networks, loop networks)
6. Pump application and analysis

B. Hydraulics – Open Channel
1. Open-channel flow (e.g., Manning's equation)
2. Culvert design
3. Spillway capacity
4. Energy dissipation (e.g., hydraulic jump, velocity control)
5. Stormwater collection (e.g., stormwater inlets, gutter flow, street flow, storm sewer pipes)
6. Flood plains/floodways
7. Flow measurement – open channel

C. Hydrology
1. Storm characterization (e.g., rainfall measurement and distribution)
2. Storm frequency
3. Hydrographs application
4. Rainfall intensity, duration, and frequency (IDF) curves
5. Time of concentration
6. Runoff analysis including Rational and SCS methods
7. Erosion
8. Detention/retention ponds

D. Wastewater Treatment
1. Collection systems (e.g., lift stations, sewer networks, infiltration, inflow)

E. Water Treatment
1. Hydraulic loading
2. Distribution systems

CIVIL–TRANSPORTATION Depth Exam Specifications

A competent transportation engineer should have a basic knowledge in drainage, soils, and pavement design. Culvert design and pavement design are knowledges that have not been tested previously under the current Civil exam specifications. **Beginning with the April 2010 exam, Section V of the Transportation module has been broadened to permit testing in these important transportation knowledges.**

	Approximate Percentage of PM Exam
I. Traffic Analysis	**22.5%**
A. Traffic capacity studies	
B. Traffic signals	
C. Speed studies	
D. Intersection analysis	
E. Traffic volume studies	
F. Sight distance evaluation	
G. Traffic control devices	
H. Pedestrian facilities	
I. Driver behavior and/or performance	
II. Geometric Design	**30%**
A. Horizontal curves	
B. Vertical curves	
C. Sight distance	
D. Superelevation	
E. Vertical and/or horizontal clearances	
F. Acceleration and deceleration	
G. Intersections and/or interchanges	
III. Transportation Planning	**7.5%**
A. Optimization and/or cost analysis (e.g., transportation route A or transportation route B)	
B. Traffic impact studies	
C. Capacity analysis (future conditions)	
IV. Traffic Safety	**15%**
A. Roadside clearance analysis	
B. Conflict analysis	
C. Work zone safety	
D. Accident analysis	
V. Other Topics	**25%**
A. Hydraulics	
1. Culvert design	
2. Open channel – subcritical and supercritical flow	
B. Hydrology	
1. Hydrograph development and synthetic hydrographs	
C. Engineering properties of soils and materials (e.g., index properties, identification of types of soils, suitable or unsuitable soil, boring logs)	

D. Soil mechanics analysis (e.g., soil behavior, soil classification, soil compaction)
E. Engineering economics
   1. Value engineering and costing
F. Construction operations and methods (e.g., erosion control measures, excavation/embankment)
G. Pavement structures (e.g., flexible and rigid pavement design)

CIVIL BREADTH MORNING SAMPLE QUESTIONS

CIVIL BREADTH MORNING SAMPLE QUESTIONS

101. Refer to the figure. The net excavation (yd^3) from Station 1+00 to Station 3+00 is most nearly:

(A) 160
(B) 262
(C) 390
(D) 463

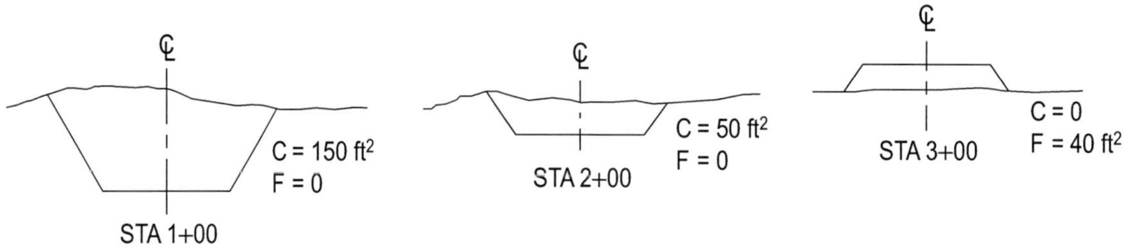

102. An existing pipe connects two maintenance holes (MH). A third MH is planned between the two. At the new MH, the elevation (ft) of the top of the pipe is most nearly:

(A) 628.56
(B) 628.06
(C) 627.56
(D) 623.06

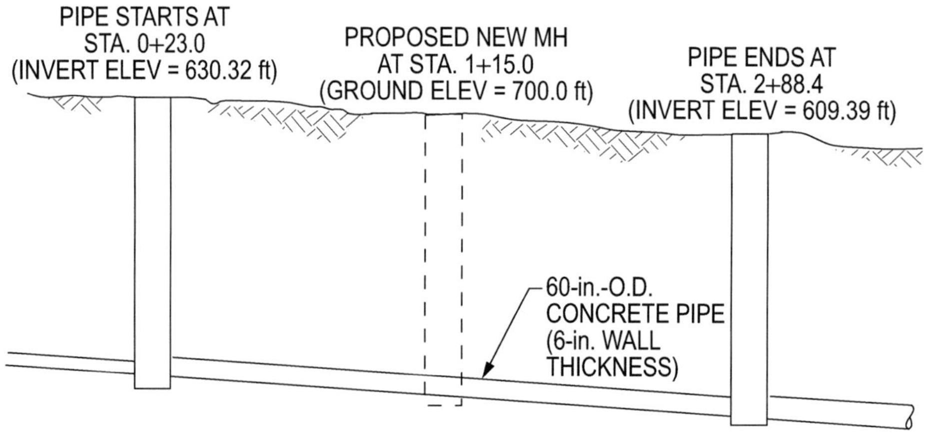

103. A 227-ft length of canal is to be lined with concrete for erosion control. With 12% waste and overbreak, the volume (yd^3) of concrete that must be delivered is most nearly:

(A) 234
(B) 280
(C) 292
(D) 327

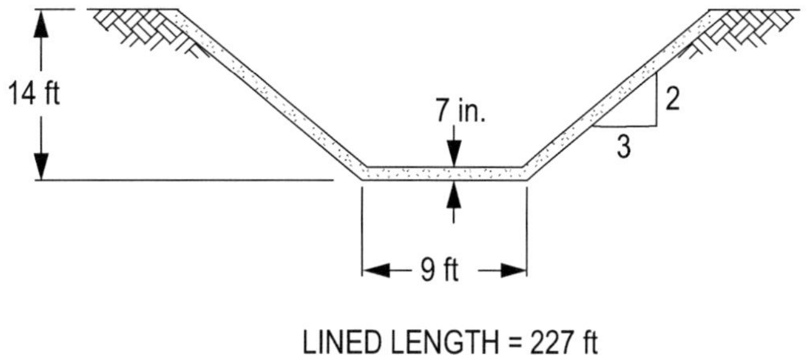

104. A track loader has an initial cost of $75,000 and a salvage value of $10,000 at the end of its expected life of 10 years. Using the straightline method of depreciation, you determine that the book value at the end of the eighth year is most nearly:

(A) $10,000
(B) $15,000
(C) $23,000
(D) $48,750

105. A CPM arrow diagram is shown below. Nine activities have been estimated with durations ranging from 5 to 35 days. The minimum time (days) required to finish the project is most nearly:

(A) 40
(B) 42
(C) 45
(D) 50

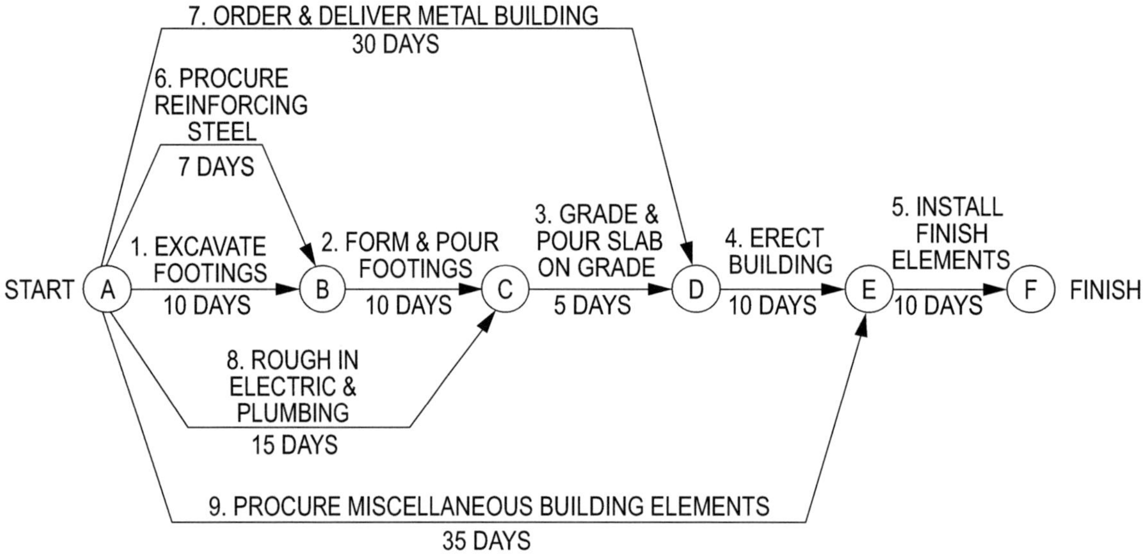

106. The budgeted labor amount for an excavation task is $4,000. The hourly labor cost is $50 per worker, and the workday is 8 hours. Two workers are assigned to excavate the material. The time (days) available for the workers to complete this task is most nearly:

(A) 3
(B) 4
(C) 5
(D) 12.5

107. If the water-to-cement ratio of concrete is decreased, which statement about the concrete is **not** true?

(A) Watertightness is decreased.

(B) Workability is decreased.

(C) Strength is increased.

(D) Durability is increased.

108. A wall form subjected to a wind load of 20 psf is prevented from overturning by diagonal braces spaced at 8 ft on center along the length of the wall form as shown in the figure. The connection at the base of the form at Point A is equivalent to a hinge. Ignore the weight of the form. The axial force (lb) resisted by the brace is most nearly:

(A) 2,050
(B) 2,560
(C) 2,900
(D) 4,525

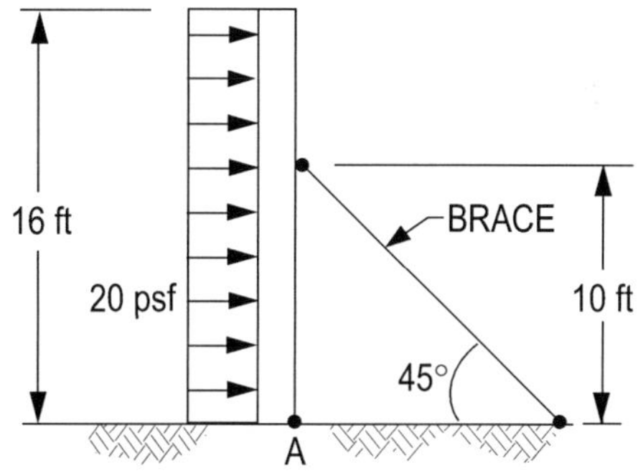

109. A sample of soil undergoes a sieve analysis and Atterberg limits testing. The results of the sieve analysis indicate that the soil is 15% gravel, 60% sand, and 25% fines with a coefficient of uniformity of 5.2 and a coefficient of curvature of 2.1. The Atterberg limits are determined to be liquid limit = 80 and plastic limit = 28. According to the Unified Soil Classification System, the soil classification is:

(A) SW

(B) SP

(C) SM

(D) SC

110. Four borrow sources are being considered for use as a sand drainage layer in a landfill cover. The materials have been classified according to the Unified Soil Classification System. Which of the following alternatives is most appropriate for the proposed use?

(A) ML

(B) SM

(C) CH

(D) SW

111. Which of the following statements regarding lateral earth pressures is correct?

(A) The lateral strain required to fully mobilize the soil passive pressure is considerably smaller than the lateral strain required to fully mobilize the soil active pressure.

(B) The lateral strain required to fully mobilize the soil passive pressure is slightly smaller than the lateral strain required to fully mobilize the soil active pressure.

(C) The lateral strain required to fully mobilize the soil passive pressure is slightly greater than the lateral strain required to fully mobilize the soil active pressure.

(D) The lateral strain required to fully mobilize the soil passive pressure is considerably greater than the lateral strain required to fully mobilize the soil active pressure.

112. A standard Proctor test was conducted with the following results:

Weight of mold plus sample	6.55 lb
Weight of mold	2.43 lb
Volume of mold	1/30 ft^3

A moisture content test was conducted with the following results:

Wet weight of soil plus tare	820 g
Dry weight of soil plus tare	793 g
Weight of tare	120 g

The dry density (pcf) and soil moisture content, respectively, are most nearly:

	Dry Density	Soil Moisture Content
(A)	118.8	4.0%
(B)	119.5	3.4%
(C)	123.6	4.0%
(D)	128.5	4.0%

113. The minimum factor of safety against rotational failure for permanent slopes under long-term, non-seismic conditions should meet or exceed:

(A) 1.0
(B) 1.1
(C) 1.5
(D) 3.0

114. A bridge footing is to be constructed in sand. The water table is at the ground surface. The ultimate bearing capacity would be based on what type of soil unit weight?

(A) Dry unit weight
(B) Saturated unit weight
(C) Buoyant unit weight
(D) Total unit weight

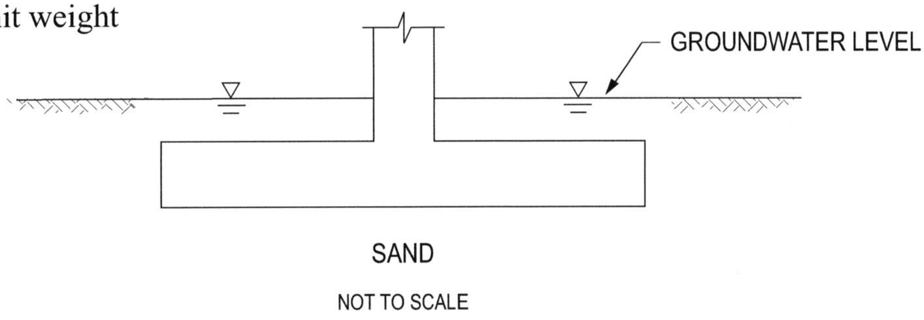

115. A sheet-pile wall retaining a silty sand is shown in the figure. Using the Rankine formula, you determine the active earth pressure coefficient is most nearly:

(A) 0.30
(B) 0.47
(C) 1.0
(D) 3.25

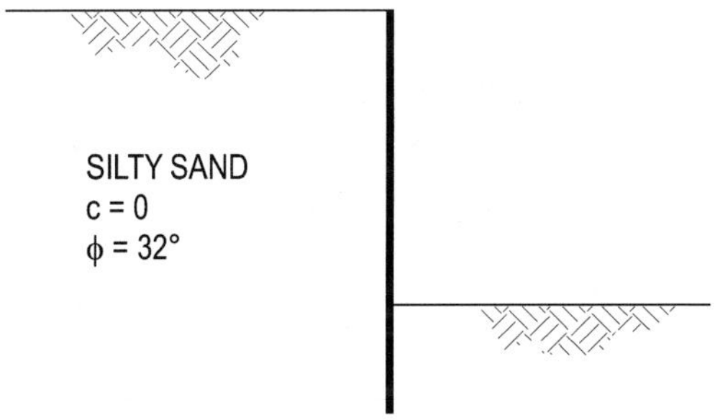

116. A masonry gravity retaining wall having a unit weight of 150 pcf is shown in the figure. Use the Rankine active earth pressure theory and neglect wall friction. The factor of safety against overturning about the toe at Point O is most nearly:

(A) 2.8
(B) 2.5
(C) 2.2
(D) 0.3

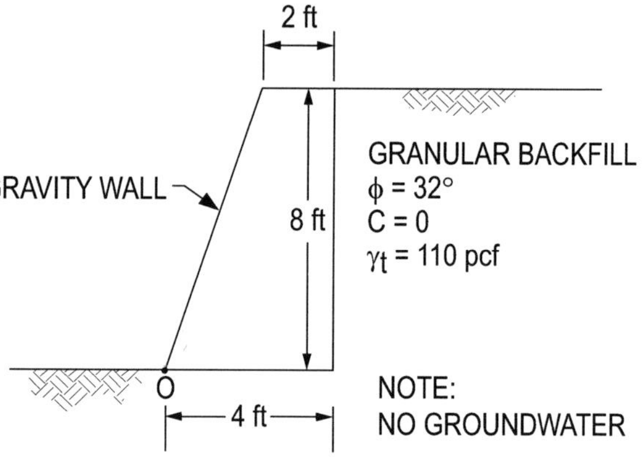

117. Referring to the figure, what load combination produces the maximum uplift on Footing A?

(A) Dead + live
(B) Dead + wind
(C) Dead
(D) Dead + live + wind

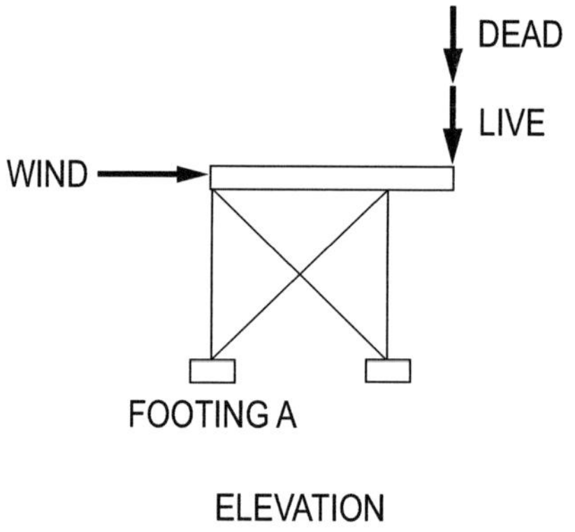

118. A simply supported truss is loaded as shown in the figure. The loads (kips) for Members b and c are most nearly:

	Member b	Member c
(A)	0	0
(B)	0	100
(C)	100	0
(D)	100	100

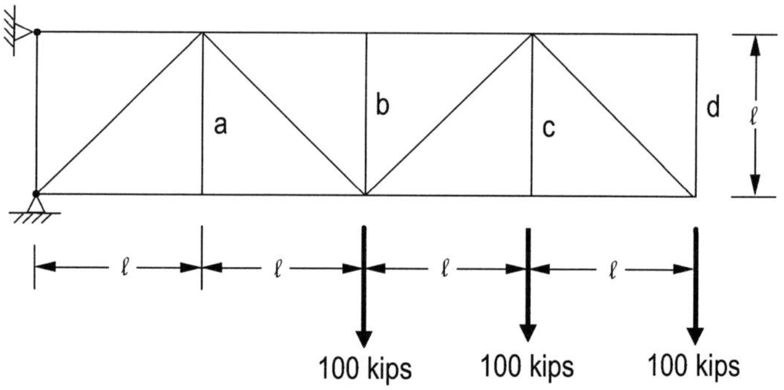

119. The beam sections shown are fabricated from 1/2-in. × 6-in. steel plates. Which of the following cross sections will provide the greatest flexural rigidity about the x-axis?

(A)

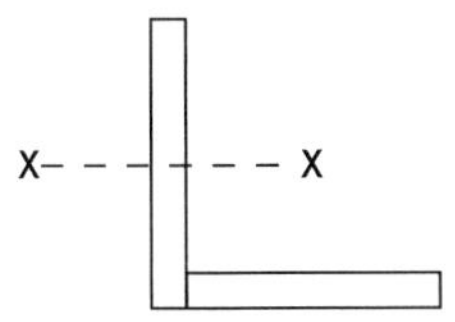

(B)

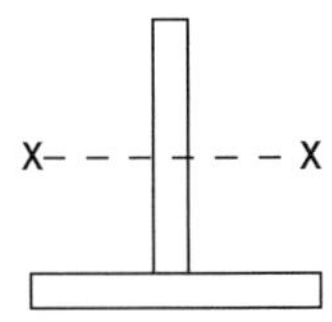

(C)

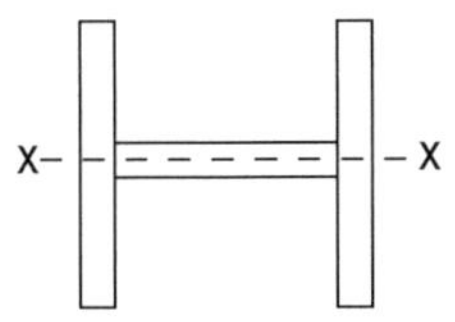

(D)

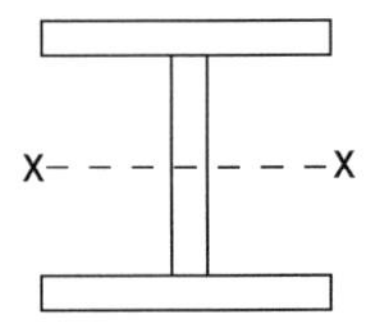

120. Which of the moment diagrams shown best represents the loading on the beam shown in the figure?

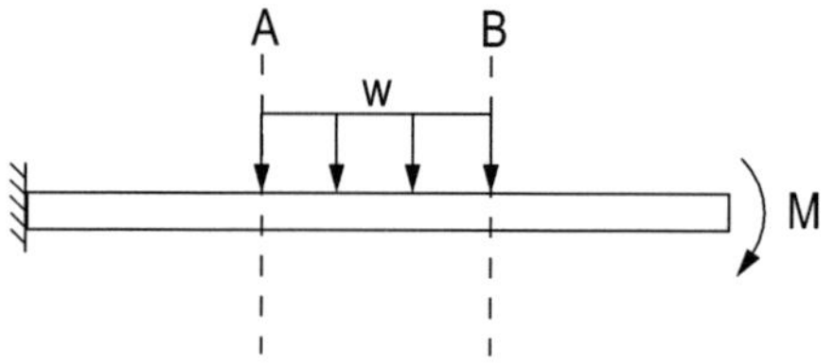

(A) 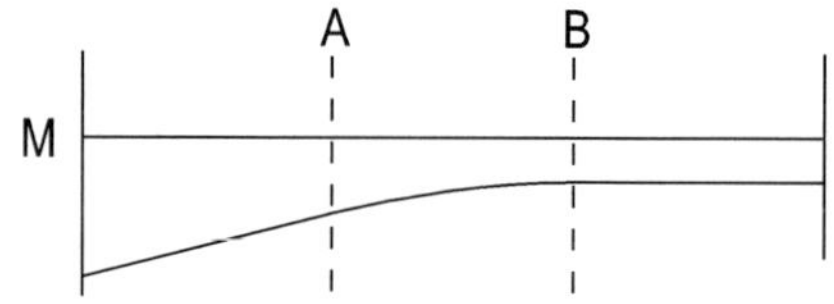

(B) 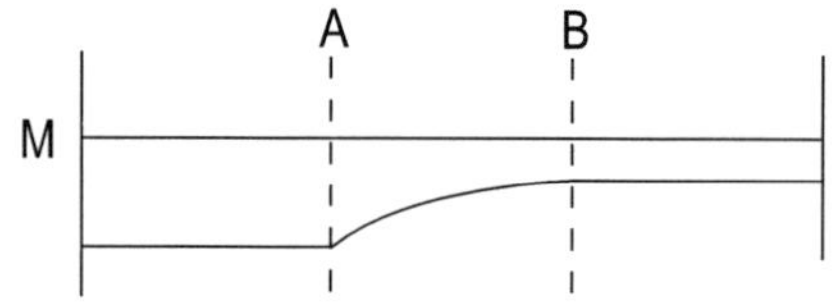

(C) 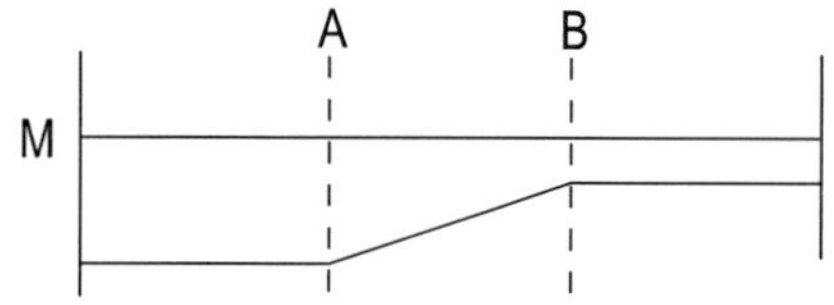

(D) 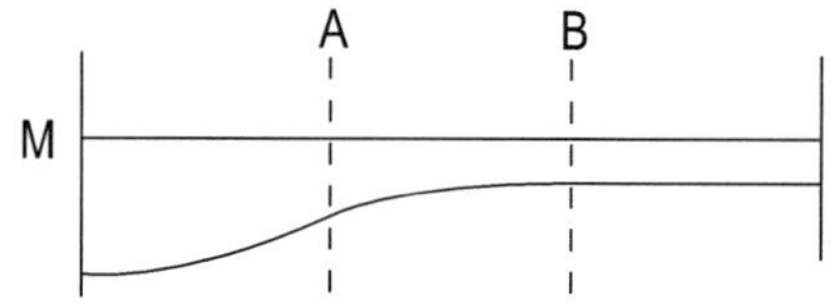

121. Consider two beams with equal cross-sections, made of the same material, having the same support conditions, and each loaded with equal uniform load per length. One beam is twice as long as the other. The maximum bending stress in the longer beam is larger by a factor of:

(A) 1.25
(B) 2
(C) 3
(D) 4

122. Sample concrete cylinders that are 6 inches in diameter and 12 inches high are tested to determine the compressive strength of the concrete f'_c. The test results are as follows:

Sample	Axial Compressive Failure Load (lb)
1	65,447
2	63,617
3	79,168

Based on the above results, the average 28-day compressive strength (psi) is most nearly:

(A) 615
(B) 2,250
(C) 2,450
(D) 2,800

123. Without taking special precautions, water used in mixing concrete must be:

(A) clean and free of substances deleterious to concrete or reinforcement

(B) non-potable

(C) deionized

(D) salty

124. The point load (kips) placed at the centerline of a 30-ft beam that produces the same maximum shear in the beam as a uniform load of 1 kip/ft is most nearly:

(A) 7.5
(B) 15
(C) 30
(D) 60

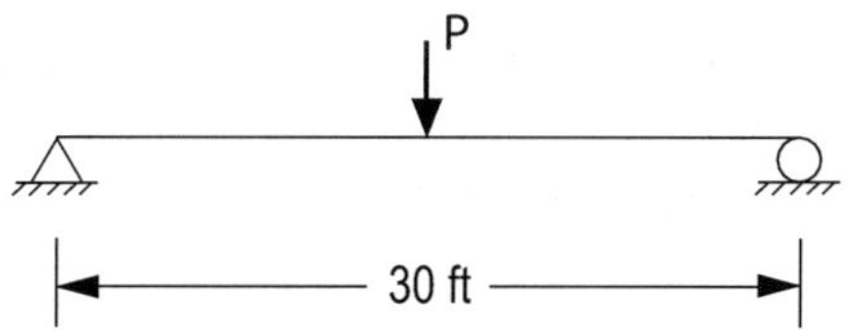

125. The following information is for a proposed horizontal curve in a new subdivision:

PI station	12+40.00
Degree of curve	10°
Deflection angle	12°30′

The station of the PT is most nearly:

(A) 12+79.80
(B) 12+80.10
(C) 13+02.00
(D) 13+64.75

126. For the sag vertical curve shown, the tangent slope at Station 14+00 is most nearly:

(A) +0.53%
(B) +1.23%
(C) +2.12%
(D) +2.77%

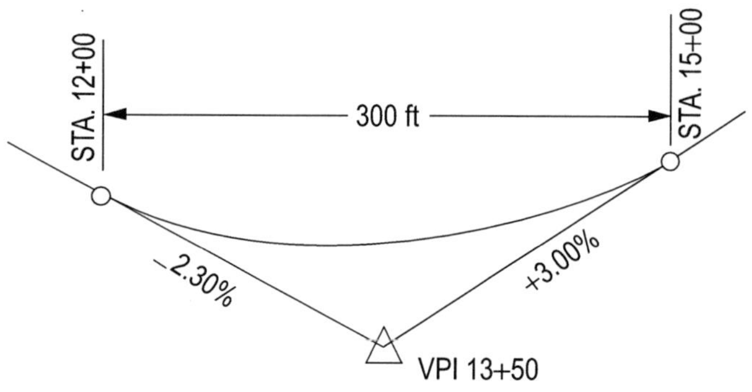

127. A driver brakes suddenly to avoid an obstacle while driving downhill on a flat grade at 35 mph on a dry pavement. Assume the deceleration rate is 11.2 ft/sec^2, and the perception-reaction time is 2 sec. The stopping distance (ft) is most nearly:

(A) 118
(B) 170
(C) 213
(D) 221

128. The figure shows a road passing near a building. Using a stopping sight distance of 700 ft, you determine the minimum distance (ft) the building can be located from the center of the interior travel lane is most nearly:

(A) 54.64
(B) 60.64
(C) 64.60
(D) 66.64

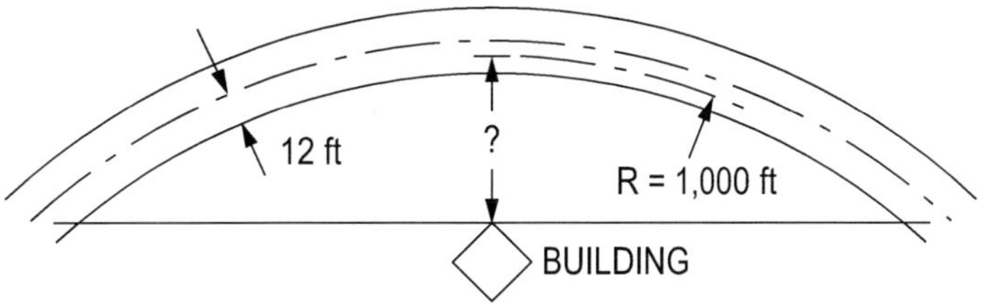

129. The vertical alignment of a section of proposed highway is shown in the figure. The vertical clearance (ft) between the bridge structure at Station 73+00 and the vertical curve is most nearly:

(A) 15.3
(B) 19.0
(C) 19.8
(D) 22.1

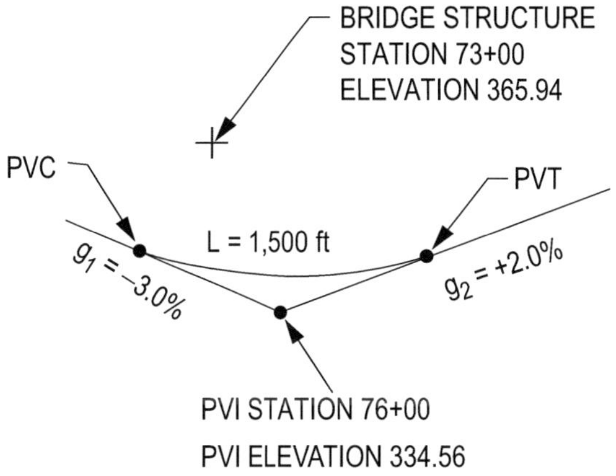

130. The purpose of superelevation is to counterbalance:

(A) lateral wind forces
(B) centripetal force
(C) gravitational force
(D) frictional force between tires and pavement

131. Which of the following interchanges includes a weaving section?

(A)

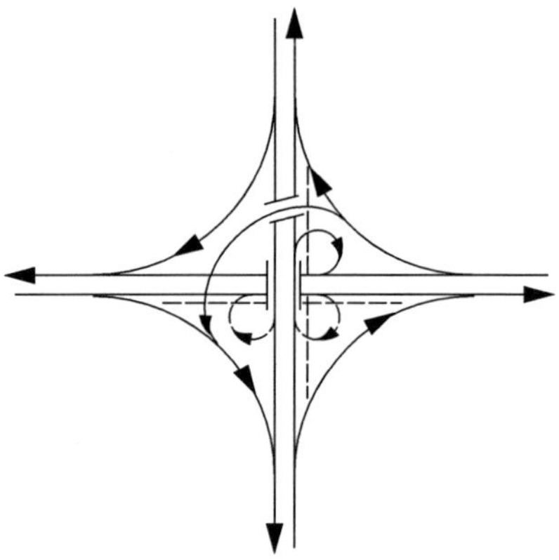

(B)

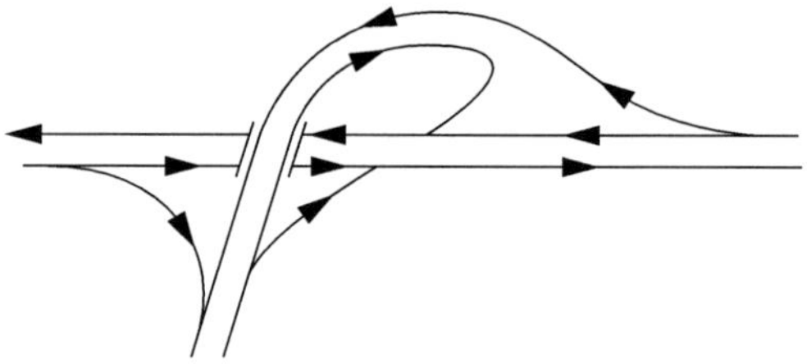

(C)

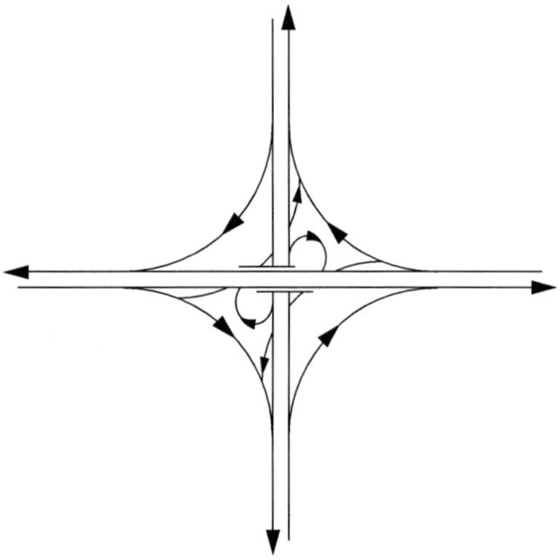

(D)

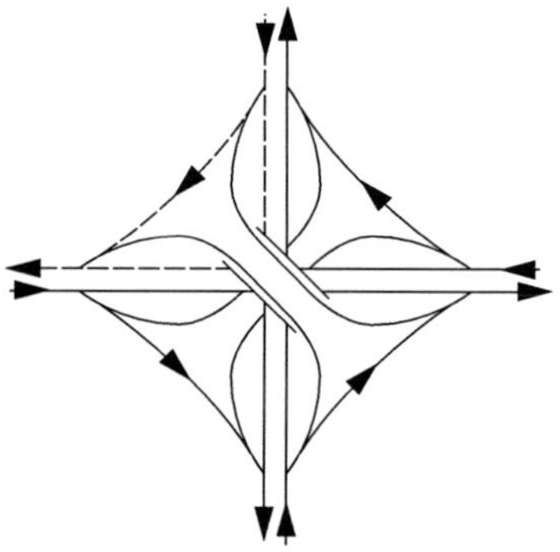

132. The difference in braking distance (ft) between a vehicle traveling at 70 mph on a level highway, and one traveling at 70 mph on a 6% downhill grade, given a deceleration rate of 11.2 ft/sec^2, is most nearly:

(A) 70
(B) 98
(C) 130
(D) 163

GO ON TO THE NEXT PAGE

133. Reservoir A and Reservoir B are connected by a siphon as shown in the figure. The difference in elevation head (ft) between the reservoirs is most nearly:

(A) 30
(B) 40
(C) 50
(D) 60

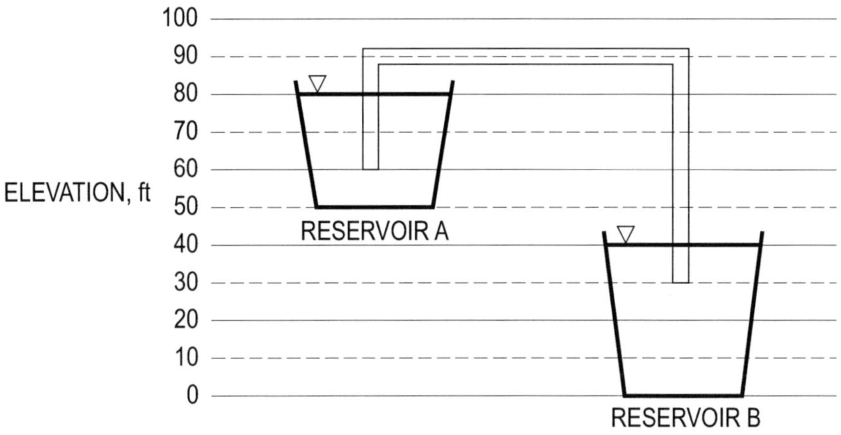

134. Assume fully turbulent flow in a 1,650-ft section of 3-ft-diameter pipe. The Darcy-Weisbach friction factor f is 0.0115. There is a 5-ft drop in the energy grade line over the section. The flow rate (cfs) is most nearly:

(A) 16
(B) 29
(C) 50
(D) 810

135. Assume a Hazen-Williams coefficient of 130. The head loss (ft) in a 3,000-ft-long, 20-in.-diameter pipeline discharging 5,000 gpm is most nearly:

(A) 5.6
(B) 7.5
(C) 12.5
(D) 14.7

GO ON TO THE NEXT PAGE

136. A drainage basin produces a stormwater runoff volume of 25.0 acre-ft, which must be drained through a rectangular channel that is 4 ft wide and 2 ft deep and has a uniform slope of 0.2%. Assume a Manning roughness coefficient of 0.022 and a constant depth of flow of 1.5 ft. The time (hr) it will take to discharge the runoff is most nearly:

(A) 12.5
(B) 16.4
(C) 18.5
(D) 25.0

137. The following table represents the rainfall recorded from all rain gages located in and around a drainage area.

Gage	A	B	C	D	E	F	G	H	I	J	K
Rainfall (in.)	2.1	3.6	1.3	1.5	2.6	6.1	5.1	4.8	4.1	2.8	3.0

Using the arithmetic mean method, you determine the average precipitation (in.) for the drainage area is most nearly:

(A) 3.4
(B) 3.7
(C) 4.1
(D) 37.0

138. You are determining the unit hydrograph from an isolated storm event. The difference between the total runoff and the base flow is the:

(A) effective rainfall intensity
(B) runoff duration
(C) direct runoff
(D) peak runoff

GO ON TO THE NEXT PAGE

139. Consider the effect of rainfall on wastewater flow rates in combined sanitary sewers. An almost immediate significant increase in flow would most likely be caused by:

(A) sustained flow

(B) infiltration

(C) delayed inflow

(D) direct inflow

140. The ratio of maximum water demand to average water demand is called the:

(A) firefighting demand

(B) critical flow

(C) peaking factor

(D) design flow

This completes the morning session. Solutions begin on page 49.

TRANSPORTATION AFTERNOON SAMPLE QUESTIONS

501. The following information applies to a four-lane freeway.

Volume	2,400 vph (in one direction)
Peak hour factor	0.90
Base free flow speed	60 mph
5% truck traffic	
12-ft lanes	
10-ft outside shoulders	
2 mile interchange spacing	
Level terrain	
No recreational vehicles	
Commuter traffic familiar with facility	

The level of service (LOS) for the freeway is most nearly:

(A) LOS A

(B) LOS B

(C) LOS C

(D) LOS D

502. The signalized intersection shown in the figure has a pedestrian demand of 0.9 pedestrians per cycle and an average pedestrian speed of 4.0 fps. The **minimum** green time (sec) for pedestrians traveling in the North-South direction is most nearly:

(A) 22
(B) 27
(C) 29
(D) 32

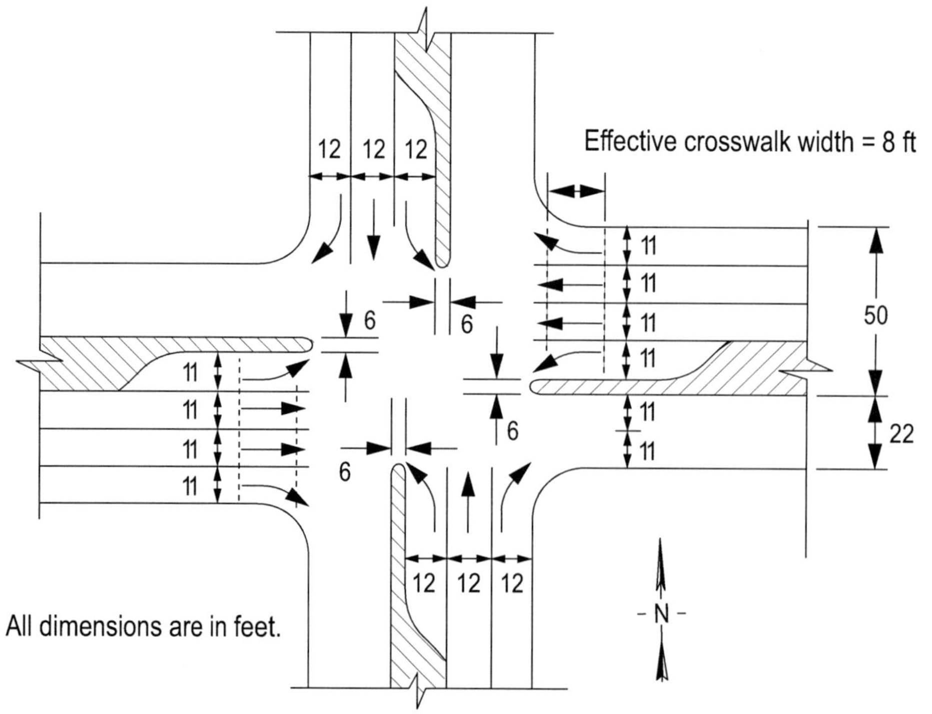

INTERSECTION PLAN
NOT TO SCALE

503. A speed study concludes that the 85th percentile speed of free flowing traffic is 56.2 mph. Which of the following posted speed limit signs most nearly meets the guidance for speed limits signs?

(A) 50 mph, 55 mph, 60 mph, 65 mph

(B) 50 mph, 55 mph

(C) 55 mph, 60 mph

(D) 60 mph, 65 mph

504. A section of a major street—an urban arterial with six traffic signalized intersections—is shown below. The traffic signals at each intersection have a total cycle length of 90 sec with 50 sec of green (and yellow) time on the major street. It is required that a progressive signal system having a maximum bandwidth of green for movement from the 1st Avenue intersection toward the 6th Avenue intersection with a progressive speed of 30 mph be designed. The offset (sec) at 6th Avenue is most nearly:

(A) 36
(B) 43
(C) 50
(D) 63

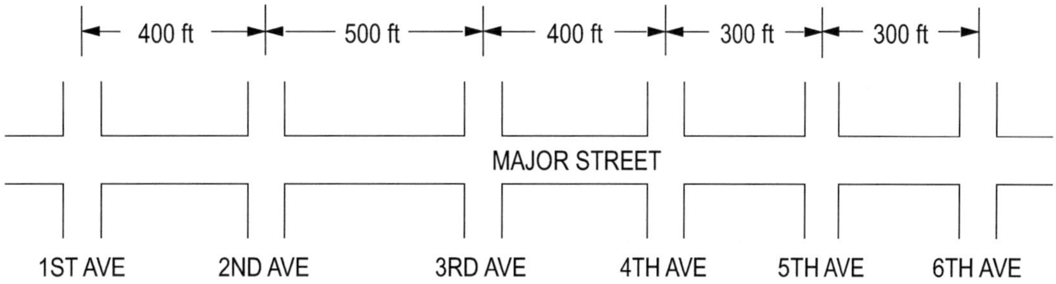

505. Continuous daily traffic counts on an urban arterial street (Street A) were recorded as shown in the following table:

Daily Traffic Counts on Urban Arterial Street A	
Day	**24-hour Volume**
Sunday	6,950
Monday	9,450
Tuesday	7,340
Wednesday	10,300
Thursday	9,850
Friday	11,250
Saturday	8,450

On a similar arterial street (Street B) in the same city, a total of 7,450 vehicles were counted in a 24-hour period on a Wednesday. The estimated average daily traffic volume for Street B for the week is most nearly:

(A) 6,570
(B) 7,080
(C) 7,450
(D) 8,450

506. If the vertical curve shown has a design speed of 60 mph, the minimum length of curve (ft) required to provide adequate stopping sight distance is most nearly:

(A) 150
(B) 420
(C) 510
(D) 570

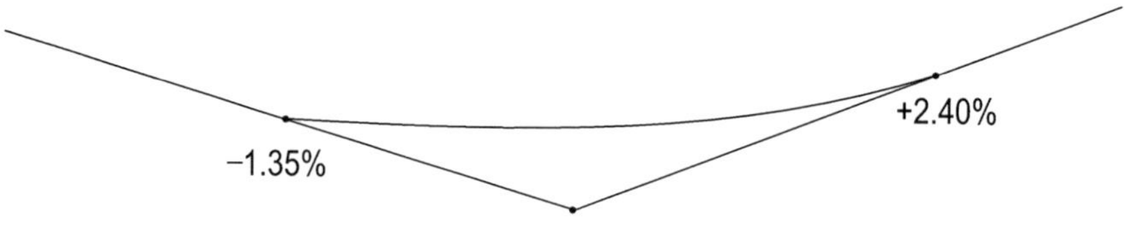

507. Delineators are to be placed on the outside of a horizontal roadway curve of 5°. The approximate spacing (ft) for the delineators along the curve is most nearly:

(A) 50
(B) 90
(C) 95
(D) 100

508. A 10-ft-wide sidewalk has an effective walkway width of 6.5 ft. The peak 15-min pedestrian flow is 1,200 pedestrians. The platoon-adjusted LOS is most nearly:

(A) LOS B
(B) LOS C
(C) LOS D
(D) LOS E

509. The following data are for a two-lane roadway:

Roadway:
24 ft wide
4-ft-wide shoulders
Rolling terrain

Traffic:
9,500 vehicles per day
Peak hour volume 11%
Peak hour factor 0.9

The peak flow rate (vehicles per hour) is most nearly:

(A) 395
(B) 565
(C) 1,045
(D) 1,160

510. A horizontal curve will be designed for a high-speed urban street having a design speed of 55 mph and a maximum superelevation rate of 4%. Using AASHTO criteria, you determine the minimum required radius (ft) is most nearly:

(A) 5,000
(B) 1,550
(C) 1,200
(D) 150

511. An existing highway has a historic site located at the PT of an existing curve with a PI located at Station 80+00, an external angle of 65°, and a radius of 1,200 ft. The curve must be relocated, moving the PI to Station 78+00. The external angle and the PC will remain the same. The radius (ft) of the new curve is most nearly:

(A) 263
(B) 314
(C) 886
(D) 1,200

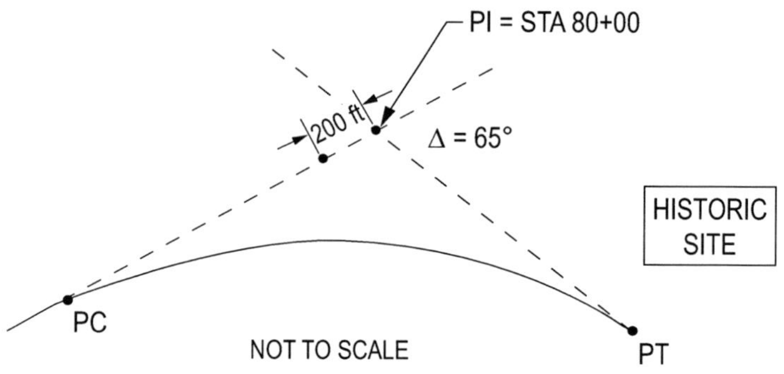

512. Two highways that have tangent horizontal alignments intersect as shown in the plan view. Each highway has a vertical alignment that consists of a vertical curve in the vicinity of the intersection point. Pertinent data for the A-2 vertical curve are as follows:

PVI station	67+00
PVI elevation	103.00 ft
Length of vertical curve	1,000 ft
Tangent grades:	
g_1	–2.0%
g_2	+3.0%

The elevation (ft) on Alignment A-2 at the intersection of these two highways is most nearly:

(A) 113.50
(B) 112.00
(C) 109.25
(D) 103.80

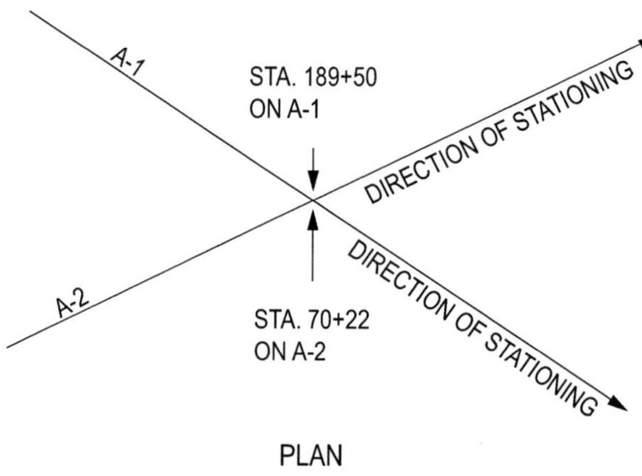

513. According to AASHTO, the passing sight distance (ft) for the following vertical curve is most nearly:

(A) 1,285
(B) 1,130
(C) 780
(D) 680

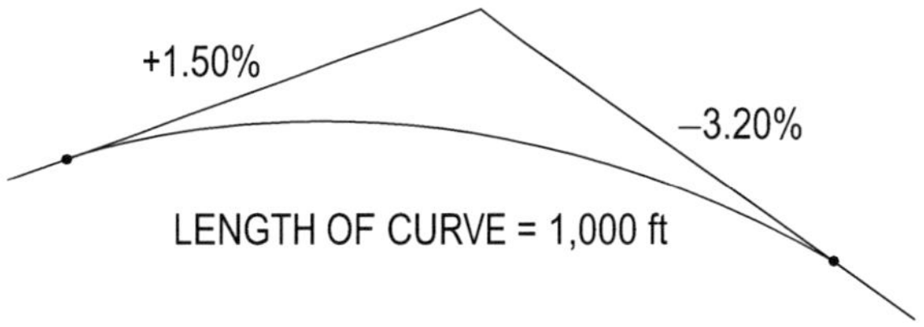

514. A horizontal curve on a two-lane rural highway has the following characteristics:

Design speed, V	60 mph
Radius (minimum)	1,091 ft
Coefficient of side friction	0.12
Lane width	12 ft

The rate of superelevation required for this curve is most nearly:

(A) 7%
(B) 10%
(C) 11%
(D) 33%

515. The **minimum** horizontal clear-zone width (ft) for a low-speed rural collector and rural local road is most nearly:

(A) 24
(B) 18
(C) 14
(D) 10

516. A single-lane entrance ramp joins a tangent section of freeway mainline as a parallel-type entrance. The entrance ramp design speed is 30 mph, and the highway design speed is 70 mph. The grade is +1.0%. The **minimum** acceleration length L (ft) needed for the entrance is most nearly:

(A) 110
(B) 520
(C) 1,230
(D) 1,350

517. Which of the following are typical characteristics of a modern roundabout in the United States:

I. Yield on entry
II. Clockwise circulation
III. Deflection of entering traffic
IV. Central island

(A) I, II, III

(B) I, II, IV

(C) I, III, IV

(D) II, III, IV

518. A vehicle is approaching a train intersection at 45 mph as shown in the figure. Assume AASHTO-recommended design values for perception-reaction time. The stop line is located 15 ft from the nearside rail, and the driver is located 8 ft back from the front bumper of the vehicle. The required sight triangle distance (ft) along the highway for a vehicle to stop at the stop line for an approaching train is most nearly:

(A) 295
(B) 347
(C) 383
(D) 438

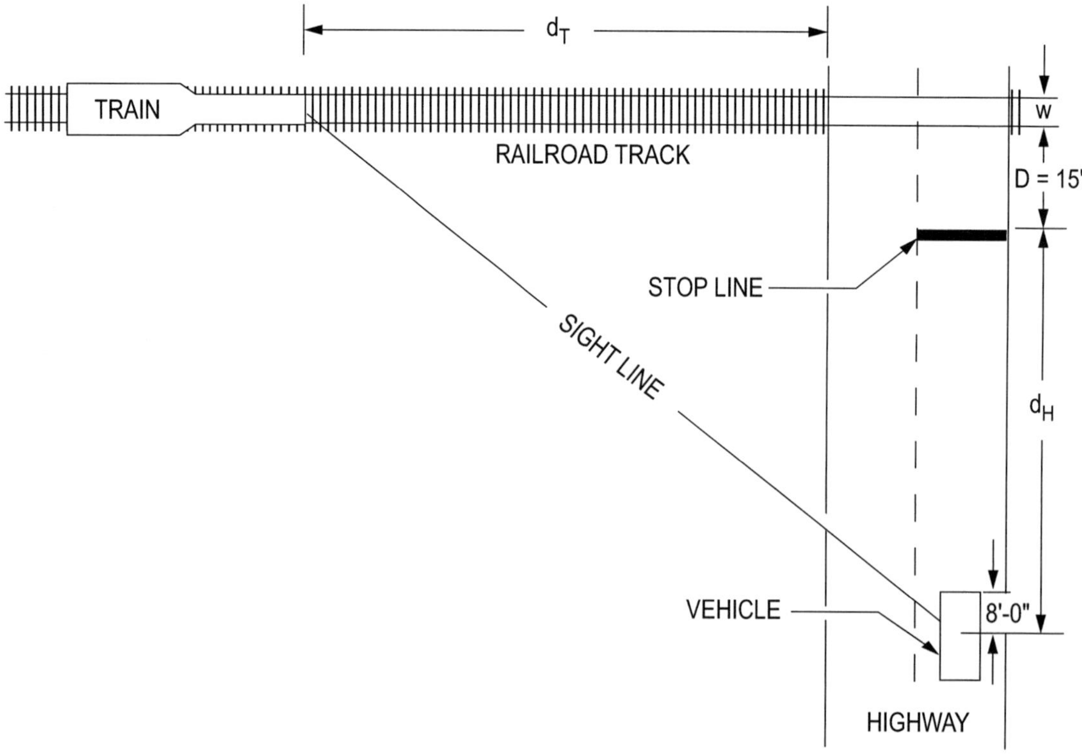

519. A section of a four-lane undivided highway forms a horizontal curve with a radius of 800 ft for the centerline of the inside lane. The highway is flat at this section and has a design speed of 45 mph. An architectural wall is being constructed by a local community center as shown in the figure. Assume a brake reaction time of 2.5 sec and a deceleration rate of 11.2 ft/sec^2. The minimum distance (ft) that the architectural wall can be constructed from the centerline of the inside lane of the curve based on sight distance is most nearly:

(A) 15
(B) 21
(C) 28
(D) 38

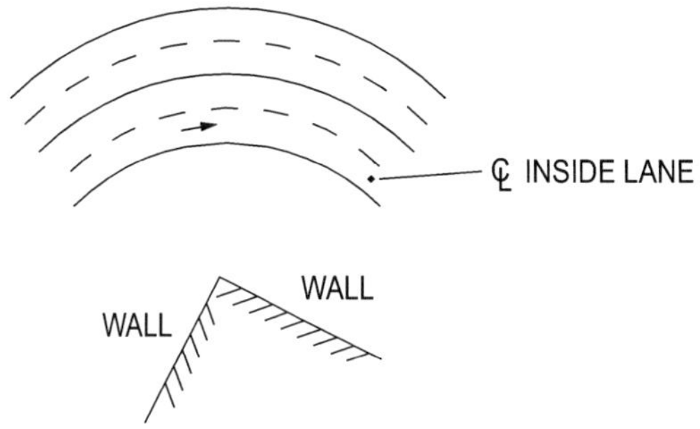

520. The tangent vertical alignment of a section of proposed highway is shown in the figure. The station of the high point is most nearly:

(A) 35+00
(B) 42+00
(C) 43+40
(D) 45+15

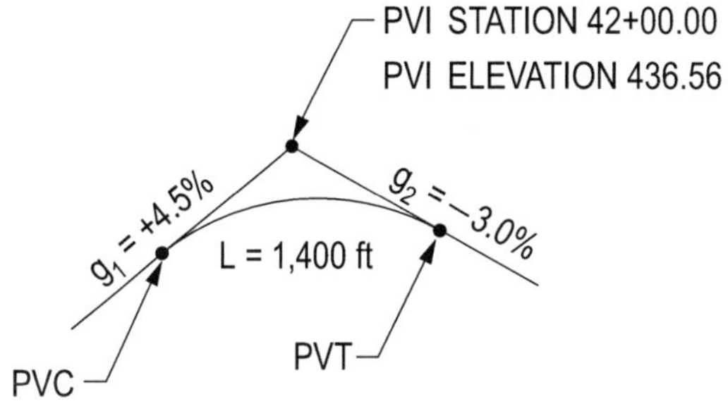

521. A freeway ramp has the following characteristics:

Radius on inner edge of pavement equals 100 ft.
One-lane, one-way operation with no provision for passing a stalled vehicle
Design traffic Condition A
Barrier curb on both sides

The ramp pavement width needed (ft) is most nearly:

(A) 15
(B) 16
(C) 17
(D) 18

522. The total cost to construct a new highway is \$6,987,500. The annual debt service to retire a bond in this amount over a 25-year period at 4% interest is most nearly:

(A) \$290,680
(B) \$363,350
(C) \$372,550
(D) \$447,270

523. A multiple regression analysis for trip generation purposes shows the following relationship for the number of trips per household:

Trips generated per household per day = 0.58 + 1.5P + 2.2A

where:

P = number of persons per household
A = number of autos per household

If a zone under study contains 600 households with an average of 4.1 persons and 2.3 autos for each household, the number of trips generated per day is most nearly:

(A) 7,100
(B) 6,700
(C) 6,600
(D) 6,200

524. A segment of a roadway has an average daily traffic of 12,350 and an annual growth rate of 7%. If this rate of traffic volume growth continues, the average daily traffic volume (vpd) in 10 years will be most nearly:

(A) 8,650
(B) 17,650
(C) 21,000
(D) 24,300

GO ON TO THE NEXT PAGE

525. The ADT is 8,000 vpd on a two-lane highway with a design speed of 60 mph. A culvert headwall is located as shown in the figure. A decision has been made to relocate the culvert headwall to achieve the AASHTO-recommended minimum clear zone. The required lateral relocation distance (ft) of the culvert headwall from its original location to achieve the AASHTO-recommended minimum clear zone is most nearly:

(A) 8
(B) 12
(C) 17
(D) 20

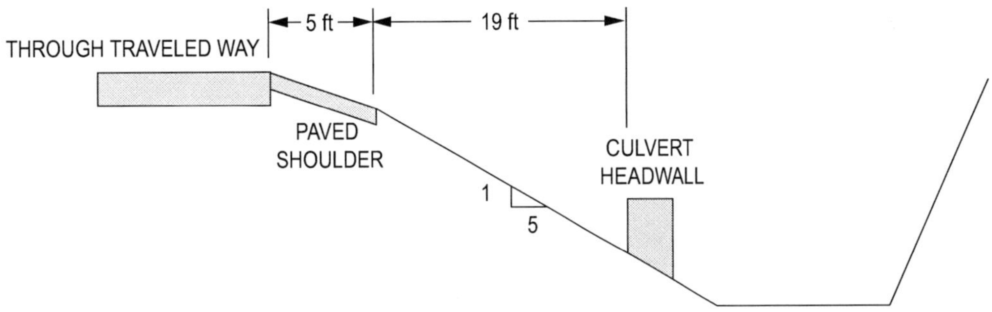

526. Which of the following traffic conflicts is most likely to increase with the installation of a traffic signal at an existing four-leg intersection?

(A) Same direction through traffic
(B) Vehicle and pedestrian in crosswalk
(C) Major street through vehicle and minor street right-turn
(D) Left-turn and opposing through vehicle

527. A freeway with a work zone speed limit of 55 mph and 12-ft lanes requires a work zone lane shift as shown. The minimum recommended length (ft) for the shifting taper is most nearly:

(A) 220
(B) 305
(C) 330
(D) 660

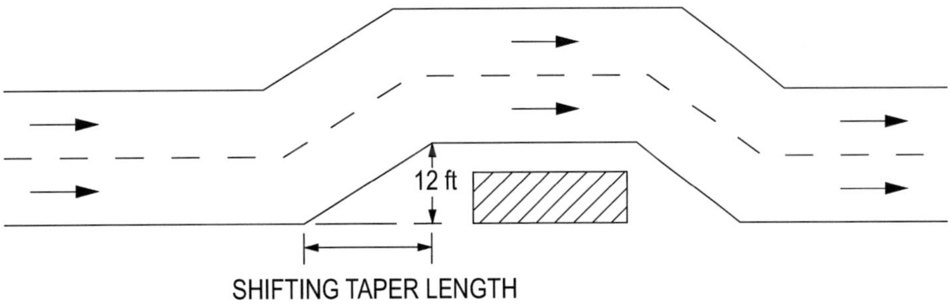

528. A vehicle ran through a stop sign and collided with a tree at an estimated speed of 30 mph. Skid marks were visible beginning at a point 150 ft in advance of and continuing through the intersection, as shown in the figure. Assume a flat grade. The estimated speed (mph) when braking began was most nearly:

(A) 86
(B) 71
(C) 68
(D) 64

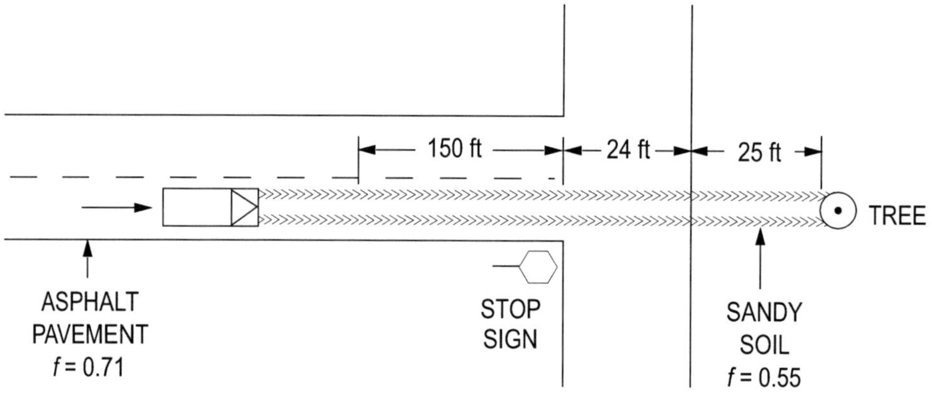

529. As part of an annual safety study, the following data was collected at four sites on rural highways with similar conditions as shown in the table.

Site	Length (miles)	Annual Number of Accidents	ADT
A	5.2	12	6,100
B	10.1	15	7,500
C	6.3	8	5,200
D	8.4	10	6,900

Assume that safety improvement funds are available for only one site and that improvement costs at each site are approximately equal. The site selected would most likely be:

(A) Site A
(B) Site B
(C) Site C
(D) Site D

530. An intersection had 25 reported traffic accidents during the months of January through September. The ADT for this intersection is shown in the following figure. The accident rate per million entering vehicles (RMEV) for this intersection is most nearly:

(A) 4.6
(B) 3.4
(C) 0.5
(D) 0.2

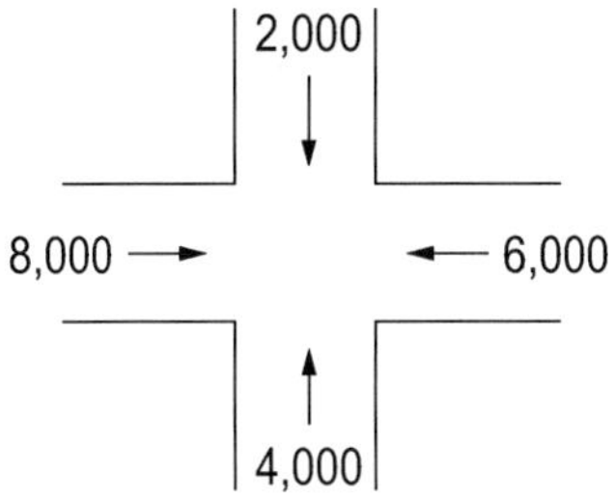

531. Runoff from a subdivision is collected in a detention basin and is released slowly through a circular corrugated metal pipe (n = 0.025) at a controlled rate of 43 cfs. The pipe slope is 0.001 ft/ft. The diameter (ft) of metal pipe that should be used to deliver the required discharge is most nearly:

(A) 3
(B) 5
(C) 6
(D) 18

532. A new rectangular open channel needs to pass 1,800 cfs without exceeding a depth of 4 ft. To ease construction, an even-numbered width is desired. Assume $n = 0.012$ and slope = 6%. The minimum channel width (ft) is most nearly:

(A) 6
(B) 10
(C) 12
(D) 16

GO ON TO THE NEXT PAGE

533. An existing watershed has the following characteristics:

Land Use	Area (acres)	*C* Value
Forest	20	0.20
Meadow	10	0.35
Residential	18	0.50
Commercial	2	0.65

The existing average *C* value for the total watershed is 0.36. A 10-acre commercial development has been proposed within the forested land. The existing time of concentration is 0.5 hour. The increase in impervious surfaces from the proposed commercial development will decrease the time of concentration by 30%. Based on the Rational formula, the increase in peak runoff from the watershed (cfs) for a 10-year storm due to the commercial development is most nearly:

(A) 14
(B) 33
(C) 43
(D) 92

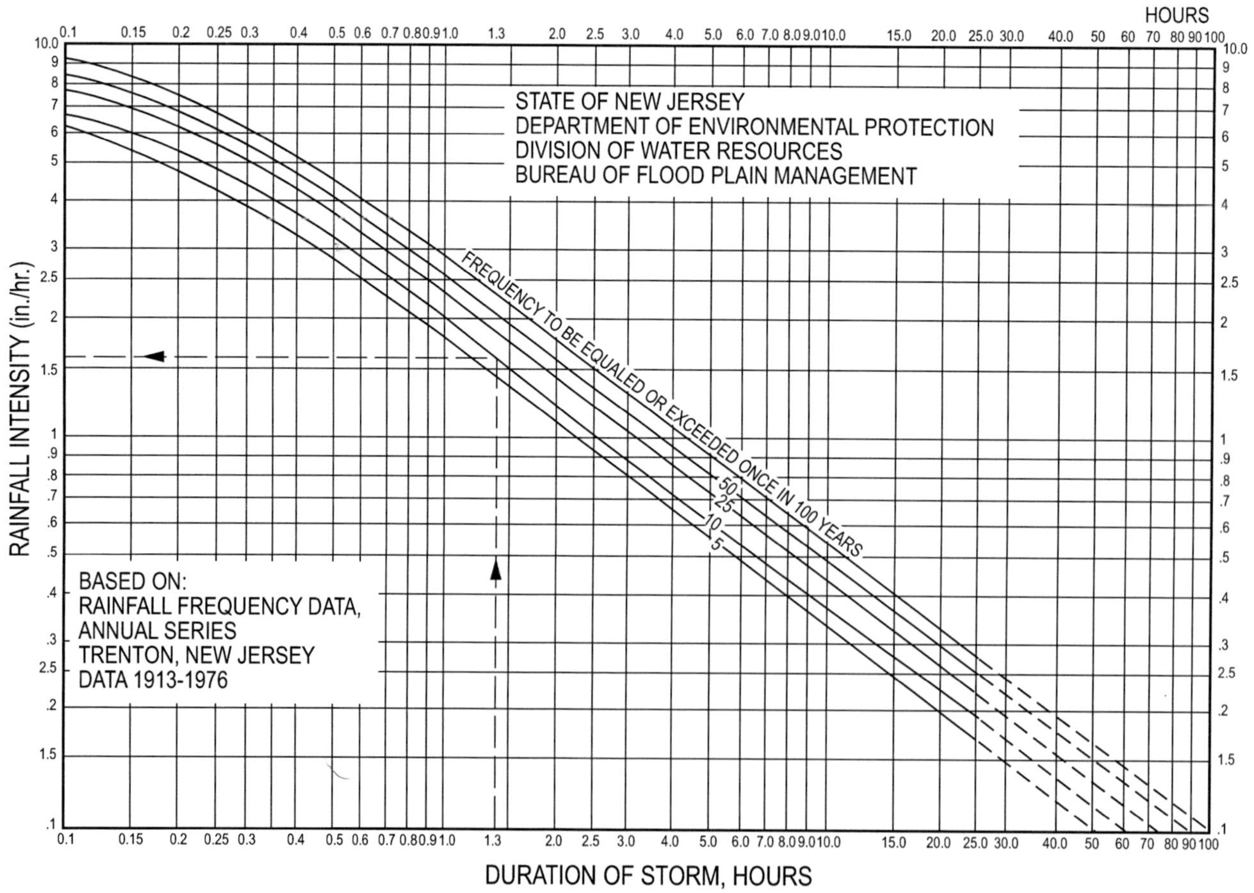

New Jersey Department of Environmental Protection, Division of Water Resources, Bureau of Flood Plain Management, 1976.

534. The following gradation data are obtained from a mechanical sieve analysis of a soil sample:

U.S. Standard Sieve Size	Percent Passing
No. 4	72%
No. 10	51%
No. 40	30%
No. 200	18%

Assume the fines are nonplastic. Based on the Unified Soil Classification System criteria, the most appropriate classification for this sample is:

(A) silty gravel

(B) well-graded clean gravel

(C) silty sand

(D) well-graded clean sand

535. The weight of a specimen of soil is 338 g, and the dried weight is 228 g. The volume of the soil before drying is 209 c.c. If the specific gravity of the soil is 2.75, the void ratio is most nearly:

(A) 1.52
(B) 1.33
(C) 0.60
(D) 0.19

GO ON TO THE NEXT PAGE

536. Based on the borehole log shown, the standard penetration test N value (blow count) for Sample 7 is most nearly:

(A) 6
(B) 8
(C) 11
(D) 14

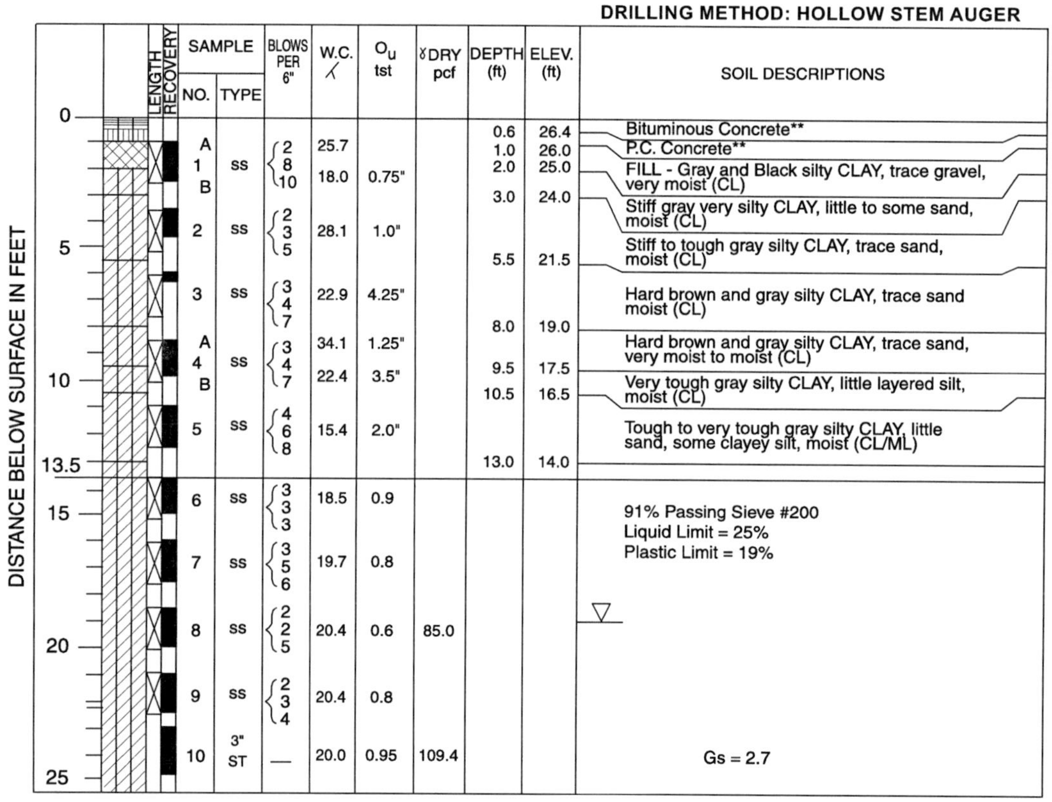

537. The Federal Highway Administration uses the following empirical relationship to determine the groundwater infiltration rate for pavements:

$$q_i = I_c\left[\frac{N_c}{W} + \frac{W_c}{W\,C_s}\right] + K_p$$

where:

q_i = design infiltration rate (ft^3/day/ft^2)
I_c = crack infiltration rate (ft^3/day/ft of crack) = 2.4
N_c = number of contributing cracks = number of lanes + 1
W_c = length of contributing transverse cracks (ft)
W = width of granular base for lane and shoulder subjected to infiltration (ft)
C_s = spacing of the transverse cracks or joints (ft)
K_p = rate of infiltration (ft^3/day/ft^2)

A two-lane roadway has the following properties:

Lane width	12 ft
Shoulder width	6 ft
W_c	20 ft
K_p	0.05 ft^3/day/ft^2
C_s	30 ft

The infiltration rate (ft^3/day/ft^2) for this pavement is most nearly:

(A) 0.23
(B) 0.29
(C) 0.34
(D) 0.43

538. Given the earthwork volumes shown in the table, find the first balance point on the mass haul diagram.

(A) Station 5+00

(B) Station 6+00

(C) Station 8+00

(D) Station 10+00

Station (ft)	Excavation (yd^3)	Embankment (yd^3)
0+00	0	0
1+00	3,000	1,000
2+00	1,000	500
3+00	1,500	1,000
4+00	2,000	500
5+00	1,000	1,000
6+00	500	2,000
7+00	0	2,000
8+00	0	1,000
9+00	500	1,000
10+00	500	0

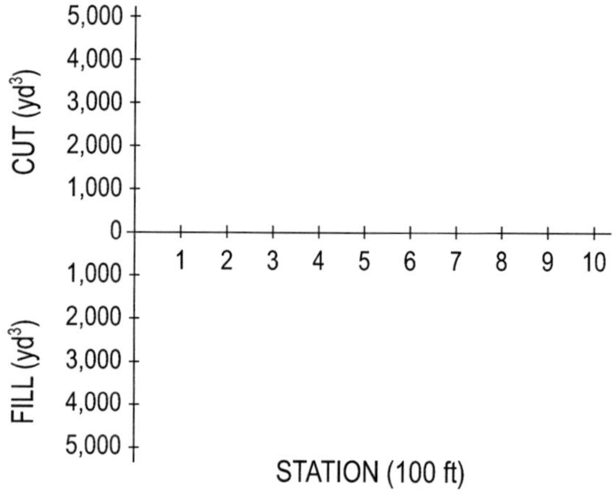

539. Given the following mass-haul diagram, the average haul distance (ft) for the section of the earthwork operation between Stations 3+00 and 9+00 is most nearly:

(A) 200
(B) 400
(C) 600
(D) 1,200

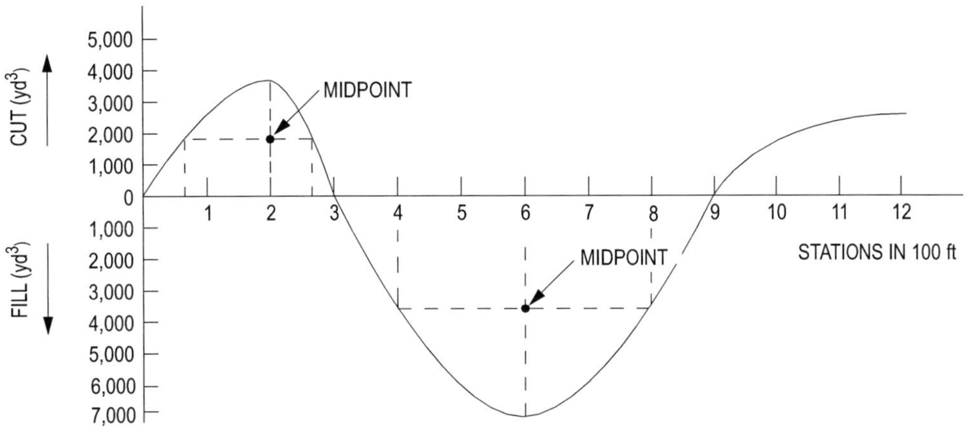

540. The following data are to be used for a flexible pavement design using the serviceability criteria:

Roadbed Moisture Condition	Season Duration (months)	Roadbed Soil Resilient Modulus, M_r (psi)
Wet	4	3,000
Dry	8	10,000

The relative damage equation is as follows:

$$u_f = (1.18 \times 10^8)(M_r^{-2.32})$$

Based on the above equation, the effective roadbed soil resilient modulus (psi) is most nearly:

(A) 7,700
(B) 6,500
(C) 4,600
(D) 2,100

This completes the afternoon session. Solutions begin on page 65.

CIVIL BREADTH MORNING SOLUTIONS

Answers to the Civil MORNING Sample Questions

Detailed solutions for each question begin on the next page.

101	C	**121**	D
102	C	**122**	C
103	D	**123**	A
104	C	**124**	C
105	D	**125**	C
106	C	**126**	B
107	A	**127**	D
108	C	**128**	B
109	D	**129**	B
110	D	**130**	B
111	D	**131**	A
112	A	**132**	B
113	C	**133**	B
114	C	**134**	C
115	A	**135**	C
116	A	**136**	C
117	D	**137**	A
118	B	**138**	C
119	D	**139**	D
120	A	**140**	C

101. Reference: Kavanagh, *Surveying with Construction Applications*, 6th ed., 2007, pp. 569–573.

Use Average End Area Method.

Stationing	Excavation (yd^3)	Embankment (yd^3)
1+00 to 2+00	$\frac{50+150}{2}\times\frac{100}{27}=370$	
2+00 to 3+00	$\frac{50+0}{2}\times\frac{100}{27}=93$	$\frac{0+40}{2}\times\frac{100}{27}=74$
Total	**463**	**74**

Net excavation = $463-74=389\ yd^3$

THE CORRECT ANSWER IS: (C)

102. Reference: Kavanagh, *Surveying with Construction Applications*, 6th ed., 2007, pp. 493–501.

Existing:

$\Delta H=(2+88.4)-(0+23.0)=288.4-23.0=265.4$ ft

$\Delta V=630.32-609.39=20.93$ ft

New:

$\Delta H=(1+15.0)-(0+23.0)=115.0-23.0=92$ ft

$$\Delta V=\frac{92}{265.4}\times 20.93=7.26 \text{ ft}$$

Inv Elev. = 630.32 – 7.26 = 623.06 ft

The top of the pipe will be above the invert elevation by (60 in. – 6 in.)/12 in./ft = 4.50 ft

623.06 + 4.50 = 627.56 ft

THE CORRECT ANSWER IS: (C)

103. Reference: Peurifoy and Oberlender, *Estimating Construction Costs,* 5th ed., Chapter 1, p. 11, Quantity Takeoff.

$$\text{Horizontal length of side slope} = 14 \times \frac{3}{2} = 21.0 \text{ ft}$$

$$\text{Slope length} = \sqrt{(14)^2 + (21)^2} = 25.24 \text{ ft}$$

$$\text{Cross-sectional area of lining} = \left[(2 \times 25.24) + 9\right]\frac{7}{12} = 34.70 \text{ ft}^2$$

$$\text{Volume of lining} = \frac{(34.70 \times 227)}{27} = 291.7 \text{ yd}^3$$

$$\text{Delivered volume} = 291.7 \text{ yd}^3 \times \underset{\text{(waste)}}{1.12} = 327 \text{ yd}^3$$

THE CORRECT ANSWER IS: (D)

104. Reference: Nunnally, *Construction Methods and Management*, 8th ed., 2011, p. 299.

$$D = \frac{\$75{,}000 - \$10{,}000}{10}$$

$$D = \$6{,}500$$

$$\text{Book value after 8 years} = \$75{,}000 - (8)(\$6{,}500) = \$23{,}000$$

THE CORRECT ANSWER IS: (C)

105. Reference: Nunnally, *Construction Methods and Management*, 8th ed., 2011, pp. 282–285.

Activities:	⑦	+	④	+	⑤		
Days:	30	+	10	+	10	=	50 days

THE CORRECT ANSWER IS: (D)

106. Reference: AGC, *Construction Planning and Scheduling*, pub. 3500.1, 6th ed., p. 37.

$$\text{Crew cost} = 2(\$50/\text{hr}) = \$100/\text{hr}$$

$$\text{Days allowed} = \frac{\$4{,}000}{(8 \text{ hr/day})(\$100/\text{hr})} = 5 \text{ days}$$

THE CORRECT ANSWER IS: (C)

107. Reference: Kosmatka, Kerkhoff, and Panarese, *Design and Control of Concrete Mixtures*, 14th ed., 2002, p. 9.

Watertightness is the ability of concrete to hold or retain water without visible leakage. Generally, less permeable concrete is more watertight. A lower water-cement ratio reduces permeability, thereby increasing water tightness.

THE CORRECT ANSWER IS: (A)

108. Reference: Hurd, *Formwork for Concrete*, ACI SP-4, 7th ed., 2005.

$$w = (20\ \text{lb/ft}^2)(8\ \text{ft}) = 160\ \text{lb/vertical ft per brace location}$$

$$\sum M_a = 0$$

$$\sum M_a = (160\ \text{lb/ft})(16\ \text{ft})(16\ \text{ft}/2) - 10\ \text{ft}\ (R_x) = 0$$

$$R_x = 2{,}048\ \text{lb}$$

$$\text{Axial load in brace} = \frac{(2{,}048)\sqrt{2}}{1} = 2{,}896\ \text{lb}$$

THE CORRECT ANSWER IS: (C)

109. SC, clayey sand with gravel, is the soil classification. The sand is the greatest quantity of the materials at 60%. The Atterberg limits plot above the A-line, putting the fines in the clay category.

THE CORRECT ANSWER IS: (D)

110. Reference: Day, *Geotechnical and Foundation Engineering*, 1999, pp. 4.44–4.45.

According to the Unified Soil Classification System, layer SW has excellent drainage properties.

THE CORRECT ANSWER IS: (D)

111. Reference: NAVFAC, DM 2.2–60.

The wall translation (or strain) required to achieve the passive state is at least twice that required to reach the active state.

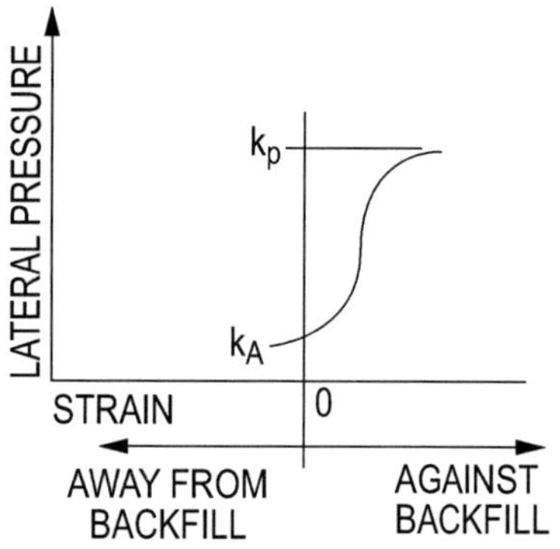

THE CORRECT ANSWER IS: (D)

112. $$\gamma_M = \frac{6.55 - 2.43}{\frac{1}{30}} = 123.6 \text{ pcf}$$

$$\text{M.C.} = \frac{820 - 793}{793 - 120} = 4.0\%$$

$$\gamma_O = \frac{123.6}{1 + 0.04} = 118.8 \text{ pcf}$$

THE CORRECT ANSWER IS: (A)

113. References: Day, *Geotechnical and Foundation Engineering*, 1999, p. 10-27, and NAVFAC 7.1-329.

The minimum factor of safety for permanent slopes is 1.5.

THE CORRECT ANSWER IS: (C)

114. The ultimate bearing capacity would be based on buoyant unit weight, also referred to as the effective unit weight.

Effective unit weight = saturated unit weight – unit weight of water

THE CORRECT ANSWER IS: (C)

115. Given the retained soil is silty sand with an internal friction angle of 32° and zero cohesion, the active earth pressure coefficient is computed using the Rankine formula as follows:

$$K_a = \tan^2\left(45 - \frac{\phi}{2}\right) = \tan^2\left(45 - \frac{32}{2}\right) = 0.307$$

THE CORRECT ANSWER IS: (A)

116. $\phi = 32°$ $\quad K_a = \tan^2(45 - \phi/2) = 0.307$

$\gamma_t = 110$ pcf $\quad P_a = (0.5)(110)(8)^2(0.307) = 1{,}081$ lb/ft

$M_a = (1{,}081)(8/3) = 2{,}883$ ft-lb/ft

$$\left.\begin{array}{l}(2)(8)(150)(1)(3) = 7{,}200 \text{ ft-lb/ft} \\ (1/2)(2)(8)(150)(1)(2/3) = 800 \text{ ft-lb/ft}\end{array}\right\} \quad \text{total} = 8{,}000 \text{ ft-lb/ft}$$

$SF = 8{,}000/2{,}883 = 2.8$

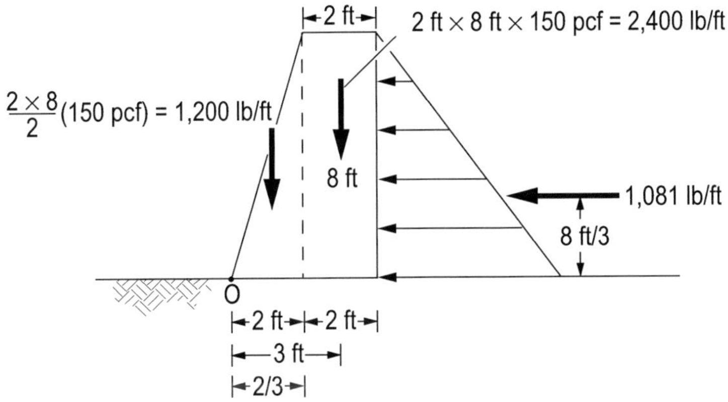

THE CORRECT ANSWER IS: (A)

117. Since the structure is cantilevered, in addition to the wind, dead load and live load will contribute to uplift.

THE CORRECT ANSWER IS: (D)

118. By inspection, Member b = 0 kips, and Member c = 100 kips.

THE CORRECT ANSWER IS: (B)

119. I_x is maximum for this section by inspection, or calculate $I_x \approx \Sigma Ad^2$ for each section.

THE CORRECT ANSWER IS: (D)

120.

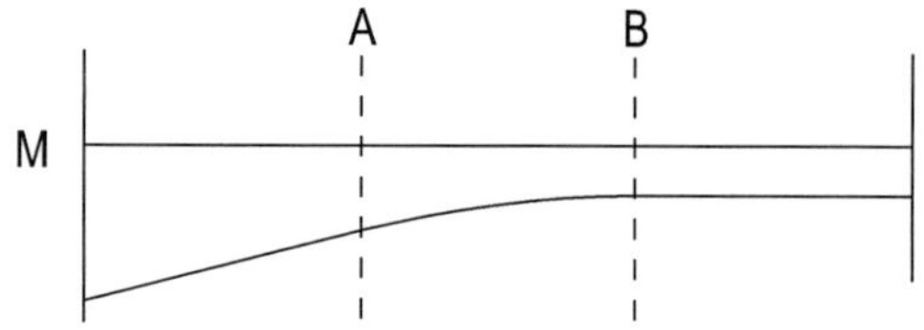

THE CORRECT ANSWER IS: (A)

121. Beam stress, $f = M/S$, where $M = wL^2/8$ and $S = bh^2/6$.
S is equal for both beams, but M varies because it depends on beam length.

Beam 1 (shorter beam): $M_1 = wL^2/8$

Beam 2 (longer beam): $M_2 = w(2L)^2/8 = 4wL^2/8$

M_2 is four times greater than M_1. Therefore the maximum bending stress is four times greater in the longer beam.

THE CORRECT ANSWER IS: (D)

122. Area $= \pi d^2/4 = 28$ in^2

Compressive stress = axial load/area

Sample 1 $\quad f'_c = \dfrac{65,447}{28} = 2,313$ psi

Sample 2 $\quad f'_c = \dfrac{63,617}{28} = 2,248$ psi

Sample 3 $\quad f'_c = \dfrac{79,168}{28} = 2,797$ psi

Average $= \dfrac{(2,313 + 2,248 + 2,797)}{3} = 2,452$ psi

THE CORRECT ANSWER IS: (C)

123. Based upon the study of concrete design, one should know water containing substances harmful to reinforcement should not be used.

THE CORRECT ANSWER IS: (A)

124. Uniform load $V = \frac{wL}{2} = \frac{1(30)}{2} = \frac{30\text{ kips}}{2} = 15\text{ kips}$

P_T load $V = \frac{P}{2} = 15\text{ kips}$

$P = 2(15) = 30\text{ kips}$

THE CORRECT ANSWER IS: (C)

125. Reference: Hickerson, *Route Location and Design*, 5th ed., p. 64.

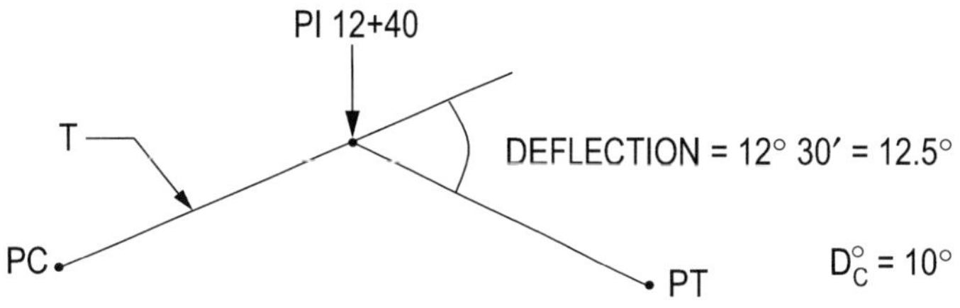

$R = 5,729.648/D_C^\circ$

$= 5,729.648/10 = 572.96\text{ ft}$

$T = R\tan\left(\frac{1}{2}\Delta\right) = R\tan(6.25°)$

$= 572.96(\tan 6.25°)$

$= 572.96(0.1095178)$

$= 62.75\text{ ft}$

Station PC = Station PI − T

$= [12+40] - 62.75$

$= 11+77.25$

Station PT = Station PC + length of curve

Length of curve $= L = 100\,\Delta/D_C^\circ$

$= 100(12.5)/10 = 125\text{ ft}$

Station PT = Station PC + 125 ft = [11 + 77.25] + 125 = 13 + 02.25

THE CORRECT ANSWER IS: (C)

126. Reference: Hickerson, *Route Location and Design*, 5th ed., pp. 154, 160.

$L = KA$
$K = L/A$
L = length of vertical curve, ft
A = algebraic difference in grades, percent $(g_2 - g_1)$
Given: VPC = 12+00
VPI = 13+50
VPT = 15+00
$g_1 = -2.30\%$
$g_2 = +3.00\%$
$L = 300$ ft

$$K = \frac{L}{A} = \frac{300}{3-(-2.3)} = 56.60 \text{ ft/percent for the vertical curve.}$$

The length from Station 14+00 to Station 15+00 = 100 ft

$$K = \frac{L}{A}$$

$$A = \frac{L}{K} = \frac{100}{56.60} = 1.77\%$$

$$A = g_2 - g_1$$

Tangent slope at Station 14+00 = g_1

$$g_1 = g_2 - A = 3.00\% - 1.77\% = 1.23\%$$

Alternate solution:

Y = elevation at a point X ft from VPC
Y' = slope at a point X ft from VPC
$X = [14+00] - [12+00] = 200$ ft
g_1 = slope 1 in ft/ft
g_2 = slope 2 in ft/ft
L = length of vertical curve, ft

$$Y = Y_{VPC} + g_1 X + \left(\frac{g_2 - g_1}{2L}\right) X^2$$

$$Y' = g_1 + \left(\frac{g_2 - g_1}{L}\right) X$$

$$Y' = -0.023 + \left(\frac{0.03-(-0.023)}{300}\right) 200 = 0.0123 \text{ ft/ft or } 1.23\%$$

THE CORRECT ANSWER IS: (B)

127. Reference: Chen and Liew, *The Civil Engineering Handbook*, 2nd ed., 1995, p. 63.8.

$$\text{Total stopping distance} = 1.47\ \text{Vt} + 1.075\left(\frac{\text{V}^2}{\text{a}}\right)$$

$$= [1.47(35)(2)] + 1.075\left(\frac{(35)^2}{11.2}\right)$$

$$= 102.9 + 117.6$$

$$= 220.5$$

THE CORRECT ANSWER IS: (D)

128. Reference: Chen and Liew, *The Civil Engineering Handbook*, 2nd ed., 1995, p. 63-16.

$$\text{Horizontal sight offset, HSO} = \text{R}\left(1 - \cos\frac{28.65\ \text{S}}{\text{R}}\right)$$

Given:

$\text{S} = 700\ \text{ft}, \quad \text{R} = 1{,}000\ \text{ft}$

$$\text{HSO} = (1{,}000)\left(1 - \cos\frac{28.65\,(700)}{1{,}000}\right)$$

$$= (1{,}000)(1 - \cos 20.055)$$

$$= (1{,}000)(1 - 0.9393639)$$

$$= (1{,}000)(0.0606361)$$

$$= 60.64\ \text{ft}$$

THE CORRECT ANSWER IS: (B)

129. Reference: Wright, *Highway Engineering*, 5th ed.

Compute the tangent elevation at Station 73+00.

$$\text{Tangent elevation} = 334.56 + 0.03(300) = 343.56 \text{ ft}$$

Compute the tangent offset, e, at the PVI station.

$$e = (L/8)(g_2 - g_1) = (15/8)\ [+2.0\% - (-3.0\%)] = 9.375 \text{ ft}$$

Compute the tangent offset, Y, at Station 73+00.

$$Y = (4e/L^2)\ X^2 = [(4)(9.375)(4.5)^2]/(15)(15) = 3.38 \text{ ft}$$

Compute the vertical curve elevation at Station 73+00.

$$\begin{aligned}\text{Curve elevation} &= \text{tangent elevation} + \text{tangent offset}\\ &= 343.56 + 3.38 = 346.94 \text{ ft}\end{aligned}$$

Compute the clearance between the bridge and the vertical curve at Station 73+00.

$$\begin{aligned}\text{Clearance} &= \text{bridge elevation} - \text{curve elevation}\\ &= 365.94 - 346.94 = 19.0 \text{ ft}\end{aligned}$$

Alternate solution:

$$Y_{VPC} = Y_{PI} + |g_1|\frac{L}{2} = 334.56 + 0.03(1{,}500/2) = 357.06$$

$$\text{VPC Station} = \text{VPI Station} - \frac{L}{2} = [76+00] - \frac{1{,}500}{2} = 68+50$$

X = distance from VPC at Station 73+00

$$X = [73{+}00] - [68{+}50] = 450 \text{ ft}$$

$$Y_x = Y_{VPC} + g_1 X + \left(\frac{g_2 - g_1}{2L}\right)X^2$$

$$= 357.06 + (-0.03)(450) + \left(\frac{0.02 - (-0.03)}{2(1{,}500)}\right)450^2$$

$$= 346.94$$

$$\begin{aligned}\text{Clearance} &= \text{bridge elevation} - \text{profile elevation}\\ &= 365.94 - 346.94 = 19.0 \text{ ft}\end{aligned}$$

THE CORRECT ANSWER IS: (B)

130. Reference: Roess, Prassas, and McShane, *Traffic Engineering,* 3rd ed., p. 48.

The summation of all forces on the vehicle travelling on a superelevated curve must be zero. The superelevated portion of the road will counterbalance the centripetal force.

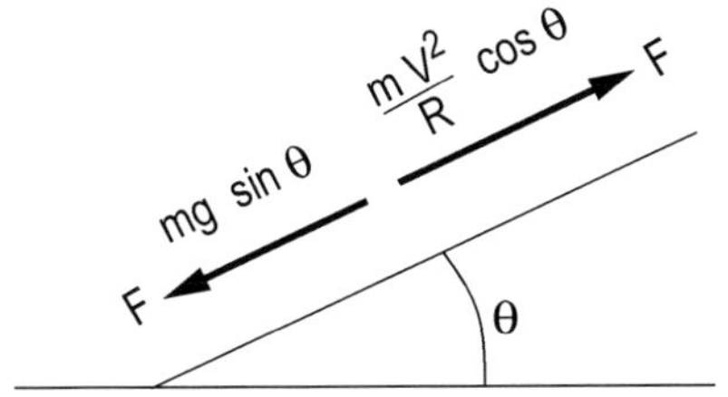

THE CORRECT ANSWER IS: (B)

131. Reference: Chen and Liew, *The Civil Engineering Handbook*, 2nd ed., 1995, p. 63-28.

The weaving sections are between loops (two cases).

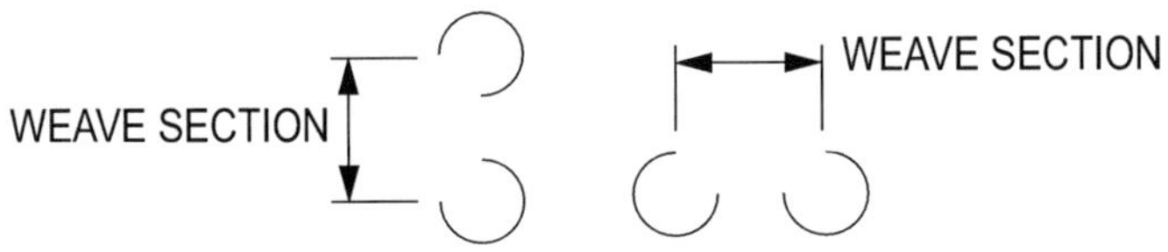

THE CORRECT ANSWER IS: (A)

132. Reference: Roess, Prassas, and McShane, *Traffic Engineering*, 3rd ed., p. 31.

$$D_{b\,0\%} = \frac{V^2}{30\left(\frac{a}{g} \pm G\right)} = \frac{70\text{ mph}^2}{30\left(\frac{11.2}{32.2}\right)} = 469.58\text{ ft}$$

$$D_{b\,6\%} = \frac{V^2}{30\left(\frac{a}{g} \pm G\right)} = \frac{70\text{ mph}^2}{30\left(\frac{11.2}{32.2} - (0.06)\right)} = 567.47\text{ ft}$$

$$\text{Difference} = D_{b\,6\%} - D_{b\,0\%} = 567.47 - 469.58 = 97.89\text{ ft}$$

THE CORRECT ANSWER IS: (B)

133. Reference: Mott, *Applied Fluid Mechanics*, 6th ed., 2006, p. 168.

The difference in elevation head is the difference in the water level elevations of the two reservoirs.

$$\Delta Z = Z_A - Z_B = 80 \text{ ft} - 40 \text{ ft}$$

$$\Delta Z = 40 \text{ ft}$$

THE CORRECT ANSWER IS: (B)

134. Reference: Merritt, Loftin, and Ricketts, *Standard Handbook for Civil Engineers*, 4th ed., 1996, pp. 21.22 and 21.42.

The Darcy-Weisbach equation is $h_f = f \dfrac{L}{D} \dfrac{V^2}{2g}$

where

h_f = headloss, ft
f = friction factor, unitless
L = length, ft
D = diameter of pipe, ft
V = velocity, ft/sec
g = gravitational constant, 32.2 ft/sec^2

Substituting gives

$$5 \text{ ft} = 0.0115 \times \frac{1{,}650 \text{ ft}}{3.0 \text{ ft}} \times \frac{V^2}{2 \times 32.2 \text{ ft/sec}^2}$$

$$V^2 = 50.91 \text{ ft}^2/\text{sec}^2$$

$$V = 7.135 \text{ ft/sec}$$

$$Q = VA = V \times \frac{\pi}{4} D^2 = 7.135 \text{ ft/sec} \times \frac{\pi}{4} (3.0 \text{ ft})^2$$

$$Q = 50 \text{ cfs}$$

THE CORRECT ANSWER IS: (C)

135. Reference: Viessman and Hammer, *Water Supply and Pollution Control*, 7th ed., 2005, p. 128.

The Hazen-Williams equation for a circular pipe is

$$Q = 0.279\, C\, D^{2.63}\, S^{0.54} = 0.279\, C\, D^{2.63} (h_L / L)^{0.54}$$

where

- Q = flow, MGD
- D = diameter, ft
- S = slope = h_L/L, unitless
- h_L = headloss, ft
- L = length, ft

$$(5{,}000 \text{ gpm})\left(\frac{1 \text{ MGD}}{694 \text{ gpm}}\right) = 0.279 \times 130 \times \left(20 \text{ in.} \times \frac{1 \text{ ft}}{12 \text{ in.}}\right)^{2.63} \left(\frac{h_L}{3{,}000 \text{ ft}}\right)^{0.54}$$

$$h_L = 12.5 \text{ ft}$$

THE CORRECT ANSWER IS: (C)

136. Reference: Mott, *Applied Fluid Mechanics*, 6th ed., 2006, p. 450.

$$Q = VA = \frac{1.49}{n} A R^{2/3} S^{1/2}$$

$$= \frac{1.49}{0.022}(1.5 \text{ ft} \times 4 \text{ ft})\left(\frac{(1.5 \text{ ft} \times 4 \text{ ft})}{4 + 2(1.5 \text{ ft})}\right)^{2/3} (0.002)^{1/2}$$

$$= 16.4 \text{ cfs}$$

$$\text{Time} = 25 \text{ acre-ft} \times \frac{43{,}560 \text{ ft}^3}{1 \text{ acre-ft}} = 1.089 \times 10^6 \text{ ft}^3$$

$$= \frac{1.089 \times 10^6 \text{ ft}^3}{16.4 \text{ ft}^3/\text{sec}} \times \frac{1 \text{ min}}{60 \text{ sec}} \times \frac{1 \text{ hr}}{60 \text{ min}}$$

$$= 18.5 \text{ hours}$$

THE CORRECT ANSWER IS: (C)

137. Reference: Mays, *Water Resources Engineering*, 2001, p. 211.

According to the arithmetic mean method, the average precipitation is simply the average of all the rainfall gages.

Average precipitation = (2.1 + 3.6 + 1.3 + 1.5 + 2.6 + 6.1 + 5.1 + 4.8 + 4.1 + 2.8 + 3.0)/11
Average precipitation = 3.4 in.

THE CORRECT ANSWER IS: (A)

138. References: Viessman and Hammer, *Water Supply and Pollution Control*, and Gupta, *Hydrology and Hydraulic Systems*, 7th ed., 2005, pp. 240–241.

Total runoff – base flow = direct runoff

THE CORRECT ANSWER IS: (C)

139. Reference: Metcalf and Eddy, *Wastewater Engineering: Treatment and Reuse*, 4th ed., p. 163.

"Direct inflow: Those types of inflow that have a direct stormwater runoff connection to the sanitary collection system and cause an almost immediate increase in wastewater flowrates."

THE CORRECT ANSWER IS: (D)

140. References: McGhee, *Water Supply and Sewerage*, 6th ed., pp. 15–16, 45, and Viessman and Hammer, *Water Supply and Pollution Control*, 7th ed., 2005, pp. 249–255.

Firefighting demand is the amount and rate of water used for fighting fires.

Critical flow defines a state of flow between two flow regimes. Critical flow coincides with minimum specific energy for a given discharge and maximum discharge for a given specific energy.

Peaking factor is a term for the ratio of maximum to average flow. In addition, the other answers have definitions unrelated to flow ratios. The correct answer is C, peaking factor.

Design flow is defined as the maximum flow that can pass through a specified structure safely.

THE CORRECT ANSWER IS: (C)

TRANSPORTATION AFTERNOON SOLUTIONS

Answers to the TRANSPORTATION Afternoon Sample Questions

Detailed solutions for each question begin on the next page.

501	C	**521**	C
502	A	**522**	D
503	C	**523**	A
504	B	**524**	D
505	A	**525**	B
506	C	**526**	A
507	D	**527**	C
508	D	**528**	B
509	D	**529**	A
510	C	**530**	A
511	C	**531**	B
512	A	**532**	B
513	C	**533**	B
514	B	**534**	C
515	D	**535**	A
516	D	**536**	C
517	C	**537**	B
518	C	**538**	C
519	B	**539**	B
520	C	**540**	C

501. Reference: Transportation Research Board, *Highway Capacity Manual* (HCM), 2000, pp. 23-4, 23-5.

Compute the free flow speed (FFS):

$$FFS = BFFS - f_{LW} - f_{LC} - f_N - f_{id}$$

where:

BFFS = 60 mph (given)

$f_{LW} = 0$ 12-ft lanes

$f_{LC} = 0$ 10-ft shoulders

$f_N = 4.5$ two lanes in each direction

$f_{id} = 0$ 2 miles spacing between interchanges

$$FFS = 60 - 4.5 = 55.5 \text{ mph}$$

Compute f_{HV}, p. 23.8, level terrain $E_T = 1.5$:

$$f_{HV} = \frac{1}{1 + P_T(E_T - 1)} = \frac{1}{1 + 0.05\,(1.5 - 1)} = 0.9756$$

Compute flow rate:

$$v_p = \frac{V}{(PHF)(N)(f_{HV})(f_p)} = \frac{2,400}{0.9 \times 2 \times 0.976 \times 1.00} = 1,366 \text{ pcphpl}$$

Compute density:

$$\text{Density} = \frac{\text{volume}}{\text{speed}} = \frac{1,366}{55.5} = 24.6 \text{ pcpmpl}$$

Find LOS:

From p. 23.3, LOS = C

THE CORRECT ANSWER IS: (C)

502. Reference: *Highway Capacity Manual*, 2000, pp. 16-5, 16-48, and 16-49.

$$G_p = 3.2 + \left(\frac{L}{S_p}\right) + \left(0.27 N_{ped}\right) \qquad (\text{for } W_E \le 10 \text{ ft})$$

$$= 3.2 + \left(\frac{72}{4}\right) + \left[(0.27)(0.9)\right]$$

$$= 21.4 \text{ sec}$$

THE CORRECT ANSWER IS: (A)

503. Reference: 2003 MUTCD, or 2009 MUTCD, Section 2B.13.

When a speed limit is to be posted, it should be within 10 km/hr or 5 mph of the 85th percentile speed of the free flowing traffic.

$$\left.\begin{array}{l} 56.2 + 5 \text{ mph} = 61.2 \\ 56.2 - 5 \text{ mph} = 51.2 \end{array}\right\} 55 \text{ or } 60 \text{ mph}$$

THE CORRECT ANSWER IS: (C)

504. Reference: ITE, *Traffic Engineering Handbook*, 5th ed.

$$\text{Offset} = t = \frac{L}{S}, \text{ where } L = \text{block length, ft}$$
$$S = \text{vehicle speed, fps}$$

$$L = 400 \text{ ft} + 500 \text{ ft} + 400 \text{ ft} + 300 \text{ ft} + 300 \text{ ft} = 1{,}900 \text{ ft}$$

$$S = 30 \times 1.47 = 44.1 \text{ fps}$$

$$t = \frac{1{,}900}{44.1} = 43.08 \text{ sec}$$

THE CORRECT ANSWER IS: (B)

505. Reference: Roess, Prassas, and McShane, *Traffic Engineering*, 3rd ed., pp. 189-193.

Given:

Vehicles on Street A on Wednesday	10,300
Vehicles on Street B on Wednesday	7,450

Calculate the ratio of vehicles on Street B on Wednesday to the vehicles on Street A on Wednesday:

$$\frac{7{,}450}{10{,}300} = 0.7233$$

Calculate the average daily traffic volume for Street B:

$$\text{ADT} = 0.7233 \times \left(\frac{6{,}950 + 9{,}450 + 7{,}340 + 10{,}300 + 9{,}850 + 11{,}250 + 8{,}450}{7} \right) = 6{,}570$$

THE CORRECT ANSWER IS: (A)

506. Reference: AASHTO, *A Policy on Geometric Design of Highways and Streets*, 2004, p. 277, Ex. 3-75.

K for 60 mph is 136.

L = KA

L = 136(2.4 + 1.35) = 510 ft

THE CORRECT ANSWER IS: (C)

507. Reference: 2003 MUTCD, Section 3D.04, or 2009 MUTCD, Section 3F.04.

Calculate the radius since the degree of curve is given.

$$R = \frac{5{,}729.58}{D} = \frac{5{,}729.58}{5} = 1{,}146 \text{ ft}$$

Since the radius is greater than the largest value shown on Table 3D-1, calculate the spacing by the formula given below Table 3D-1.

$$\text{Spacing, } S = 3\sqrt{R - 50} = (3)\sqrt{1{,}146 - 50}$$

$$= (3)\sqrt{1{,}094} = 3(33.08) = 99.22 \text{ ft}$$

THE CORRECT ANSWER IS: (D)

508. Reference: *Highway Capacity Manual*, 2000, pp. 18-4 and 18-5.

$$V_p = \frac{V_{15}}{15 \times W_E}$$

where

V_p = pedestrian unit flow rate (p/min/ft)

V_{15} = peak 15-min flow rate

W_E = effective walkway width (ft)

$$V = \frac{1{,}200}{15(6.5)} = \frac{1{,}200}{97.5} = 12.3$$

Using Exhibit 18-4 and V_p = 12.3, platoon-adjusted LOS = E.

THE CORRECT ANSWER IS: (D)

509. Reference: *Highway Capacity Manual*, 2000, pp. 7-1 and 7-2.

$$v = \frac{V}{PHF}$$

$$V = ADT \times 0.11 = 9{,}500 \times 0.11 = 1{,}045$$

$$v = \frac{1{,}045}{0.9} = 1{,}161$$

THE CORRECT ANSWER IS: (D)

510. Reference: AASHTO, *A Policy on Geometric Design of Highways and Streets*, 2004, pp. 146-147, Ex. 3-15.

$$\begin{aligned} R &= \frac{V^2}{(15)(0.01\,e_{max} + f_{max})} \\ &= \frac{(55)^2}{(15)\left[0.01(4) + 0.13\right]} \\ &= 1{,}186 \text{ ft} \end{aligned}$$

THE CORRECT ANSWER IS: (C)

511. Reference: Hickerson, *Route Location and Design*, 5th ed., pp. 64-65.

Determine existing tangent distance:

$$\begin{aligned} T_{existing} &= R \times \tan\frac{\Delta}{2} \\ &= 1{,}200 \times \tan 32.5° \\ &= 1{,}200 \times 0.637 \\ &= 764.48 \text{ ft} \end{aligned}$$

Determine new tangent:

$$T_{new} = 764.48 - 200 = 564.48 \text{ ft}$$

Determine new radius:

$$\begin{aligned} R &= \frac{T}{\tan\frac{\Delta}{2}} = \frac{564.48}{0.637} \\ &= 886.06 \text{ ft} \end{aligned}$$

THE CORRECT ANSWER IS: (C)

512. Reference: Hickerson, *Route Location and Design*, 5th ed., p. 149.

$$y = \frac{A}{200\,L}x^2 \qquad x = 7{,}200 - 7{,}022 = 178 \text{ ft}$$

$$y = \frac{(-2-3)}{200\,(1{,}000)}(178)^2 = 0.79$$

Elevation on curve at Station 70+22 = 112.66 + 0.79 = 113.45 ft

Alternate Solution:

$$\text{Sta}_{PVC} = (67{+}00) - \frac{L}{2} = (67{+}00) - \frac{1{,}000}{2}$$

$$= (62{+}00)$$

$$x = (70{+}22) - (62{+}00)$$

$$= 822 \text{ ft}$$

$$Y_{PVC} = 103.00 + 0.02\left(\frac{1{,}000}{2}\right) = 113.0$$

$$Y_x = Y_{PVC} + g_1 x + \left(\frac{g_2 - g_1}{2L}\right)x^2$$

$$= 113.00 + (-0.02)822 + \left(\frac{0.03-(-0.02)}{2{,}000}\right)822^2$$

$$= 113.45 \text{ ft}$$

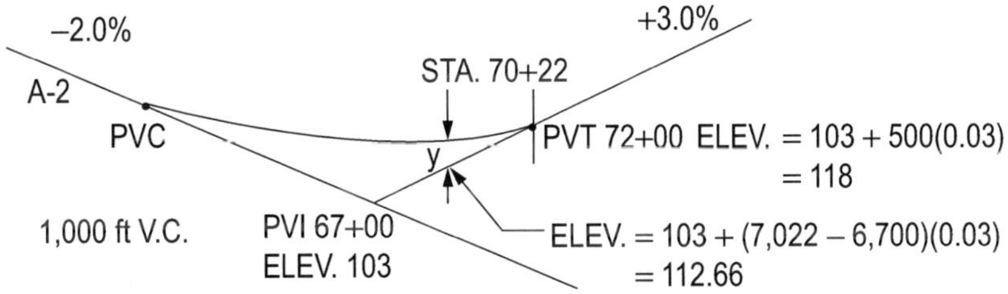

THE CORRECT ANSWER IS: (A)

513. Reference: AASHTO, *A Policy on Geometric Design of Highways and Streets*, 2004, p. 270.

Where $S < L$, $L = \dfrac{AS^2}{2{,}800}$

$$1{,}000 = \frac{[1.5 - (-3.2)](S^2)}{2{,}800}$$

$$2{,}800{,}000 = 4.7\,S^2$$

$$S^2 = 595{,}744$$

$$S = 772 \text{ ft}$$

Verify $S < L \quad \Rightarrow 772 < 1{,}000 \qquad \therefore \text{OK}$

THE CORRECT ANSWER IS: (C)

514. Reference: AASHTO, *A Policy on Geometric Design of Highways and Streets*, 2004, p. 146.

The rate of superelevation is given by the following equation:

$$0.01e + f = V^2/15R$$

where:

e	=	Rate of roadway superelevation (%)
f	=	Side friction factor
V	=	Vehicle speed (mph)
R	=	Radius of curve (ft)

$$0.01e + 0.12 = 60^2/(15)(1{,}091)$$

$$e = 10\%$$

THE CORRECT ANSWER IS: (B)

515. Reference: AASHTO, *A Policy on Geometric Design of Highways and Streets*, 2004, p. 319.

"For low-speed rural collectors and rural local roads, a minimum clear-zone width of 3.0 m (10 ft) should be provided."

THE CORRECT ANSWER IS: (D)

516. Reference: AASHTO, *A Policy on Geometric Design of Highways and Streets*, 2004.

The acceleration length given in Exhibit 10-70, p. 847, is 1,350 ft.

THE CORRECT ANSWER IS: (D)

517. Reference: AASHTO, *A Policy on Geometric Design of Highways and Streets*, 2004, pp. 574-579.

Traffic is required to yield to circulating traffic and to follow channelization (deflection) to prevent straight path through roundabout. Center islands are typical. Traffic flows **counterclockwise,** not clockwise. (I, III, IV).

THE CORRECT ANSWER IS: (C)

518. Reference: AASHTO, *A Policy on Geometric Design of Highways and Streets*, 2004, p. 735, Ex. 9-103.

$$d_H = AV_v t + \frac{\left(BV_v^2\right)}{a} + D + d_e$$

$$= (1.47)(45)(2.5) + \left(\frac{1.075(45)^2}{11.2}\right) + 15 + 8$$

$$= 383 \text{ ft}$$

THE CORRECT ANSWER IS: (C)

519. Reference: AASHTO, *A Policy on Geometric Design of Highways and Streets*, 2004, pp. 113, 227.

$$\text{Stopping sight distance} = 1.47\ Vt + 1.075\frac{V^2}{a}$$

where

t = brake reaction time, 2.5 sec

V = design speed, mph

a = deceleration rate, ft/sec^2

$$SSD = 1.47(45)(2.5) + 1.075\left[\frac{(45)^2}{11.2}\right]$$

$$= 165 + 194 = 359 \text{ ft}$$

$$HSO = R\left[1 - \cos\frac{28.65\ S}{R}\right]$$

where

S = SSD = stopping sight distance, ft

R = radius of curve, ft

HSO = horizontal sightline offset, ft

$$HSO = 800\left[1 - \cos\frac{28.65\ (359)}{800}\right]$$

$$= 800(0.025) = 20.06$$

Can also be solved graphically using Ex. 3-53.

THE CORRECT ANSWER IS: (B)

520. Reference: Hickerson, *Route Location and Design*, 5th ed., p. 149.

Compute the rate of change of grade, r.

$$r = (g_2 - g_1) / L = [-3.0\% - (+4.5\%)] / 14 \text{ sta.} = -0.5357\% / \text{sta.}$$

Compute the distance from the PVC to the high point.

$$X_{PVC} = -g_1/r = -(+4.5\%) / (-0.5357\% / \text{sta}) = 8.4002 \text{ sta}$$

Compute the PVC station.

PVC = PVI – L/2 = (42+00) – (7+00) = 35+00

Compute the station of the high point.

High point station = PVC + X_{PVC}
= (35+00) + (8+40.02)
= 43+40.02

Alternate Solution:

$$\text{Elev. at X:} \quad Y_X = Y_{PVC} + g_1 X + \left(\frac{g_2 - g_1}{2L}\right) X^2$$

$$\text{Slope at X:} \quad Y'_X = g_1 + \left(\frac{g_2 - g_1}{L}\right) X$$

At high point, slope = 0; rearrange for X

$$X = \frac{-g_1 L}{g_2 - g_1} = \frac{(-0.045)(1,400)}{(-0.03) - (0.045)} = 840.00 \text{ ft}$$

Compute PVC station: $\quad PVC = PVI - \frac{L}{2} = (42+00) - \frac{1,400}{2} = 35+00$

Compute high point station: PVC + X = (35+00) + 840 = 43+40

THE CORRECT ANSWER IS: (C)

521. Reference: AASHTO, *A Policy on Geometric Design of Highways and Streets*, 2004, p. 839 Ex. 10-67.

Determine width of freeway ramp with:

Inner edge radius = 100 ft
One-lane, one-way operating with no provision for passing stalled vehicle
Design traffic condition A
Barrier curb on both sides

Width = 15 ft + 2 ft for vertical curb on both sides
= 17 ft

THE CORRECT ANSWER IS: (C)

522. Reference: Ricketts, Loftin, and Merritt, *Standard Handbook for Civil Engineers*, 5th ed., p. 1.15.

Determine the annual debt service:

$$R = P\frac{i}{1-(1+i)^{-n}}$$

where

R = annual payment

P = present worth of investment

i = interest rate

n = number of interest periods

$$R = (\$6,987,500)\left(\frac{0.04}{1-(1+0.04)^{25}}\right)$$

$$= \$447,271.56$$

Alternatively, from economic factor tables available in various references, CRF = 0.06401

$$A = P(A/P, 4\%, 25)$$
$$= (\$6,987,500)(0.06401)$$
$$= \$447,269.88$$

THE CORRECT ANSWER IS: (D)

523. Reference: Garber and Hoel, *Traffic and Highway Engineering*, 3rd ed., 2002, pp. 520–521.

Total trips = trips per household × total households

$$= (0.58 + 1.5\,P + 2.2A) \times 600$$
$$= [0.58 + 1.5\,(4.1) + 2.2(2.3)] \times 600$$
$$= 11.79 \times 600$$
$$= 7,074$$

THE CORRECT ANSWER IS: (A)

524. Reference: Newnan, *Engineering Economic Analysis*, 3rd ed., cover page.

Given:

ADT = 12,350 vpd

Growth rate = 7%

Determine ADT 10 years in future if growth rate continues:

$$ADT_{future} = ADT_{exist}(1+0.07)^{10}$$
$$= 12,350(1.967)$$
$$= 24,294 \text{ vpd}$$

THE CORRECT ANSWER IS: (D)

525. Reference: AASHTO, *Roadside Design Guide*, 2006, p. 3-6, Table 3.1.

The minimum recommended clear zone using ADT > 6,000 vpd, 1V:5H foreslope is 36 ft. The existing condition shows the culvert headwall is located 19 ft + 5 ft = 24 ft from the edge of through travel way. Therefore, the required relocation distance to achieve the minimum recommended clear zone is 36 ft – 24 ft = 12 ft.

THE CORRECT ANSWER IS: (B)

526. Reference: 2009 MUTCD, Chapter 4B.03.

Installation of a traffic signal can result in significant increases in the frequency of collisions (especially rear-end collisions).

THE CORRECT ANSWER IS: (A)

527. Reference: 2009 MUTCD, Chapter 6C, Figure 6C-2 and Table 6C-4.

A shifting taper is at least 0.5 L. (Table 6C-3)

L = WS = 12(55) = 660 ft (Table 6C-4)

Shifting taper = 0.5 L = 330 ft minimum

THE CORRECT ANSWER IS: (C)

528. Reference: Garber and Hoel, *Traffic and Highway Engineering*, 3rd ed., 2002, pp. 64–65.

For the distance traveled on sandy soil before hitting the tree:

$$S_1 = \left(V_1^2 - V_f^2\right)/30f$$

$$25 = \left(V_1^2 - (30)^2\right)/(30 \times 0.55)$$

$$V_1 = 36.2 \text{ mph}$$

Thus, the vehicle was traveling at 36.2 mph when it left the asphalt pavement.
Now repeat these calculations to determine the speed when braking began.

$$S_2 = \left(V_2^2 - V_1^2\right)/(30 \times 0.71)$$

$$(150 + 24) = \left(V_2^2 - (36.23)^2\right)/(30 \times 0.71)$$

$$V_2 = 70.8 \text{ mph}$$

THE CORRECT ANSWER IS: (B)

529. Reference: Garber and Hoel, *Traffic and Highway Engineering*, 3rd ed., 2002, pp. 139–140.

Determine highest accident rate per hundred million vehicle miles (HMVM).

$$\text{Rate} = \frac{(\text{number of accidents})(10^8)}{(\text{ADT})(\text{ number of years})(365\text{ days/year})(L_{mi})}$$

$$\text{Rate}_A = \frac{(12)(10^8)}{(6{,}100)(1)(365)(5.2)} = 104/\text{HMVM}$$

$$\text{Rate}_B = \frac{(15)(10^8)}{(7{,}500)(1)(365)(10.1)} = 54/\text{HMVM}$$

$$\text{Rate}_C = \frac{(8)(10^8)}{(5{,}200)(1)(365)(6.3)} = 67/\text{HMVM}$$

$$\text{Rate}_D = \frac{(10)(10^8)}{(6{,}900)(1)(365)(8.4)} = 47/\text{HMVM}$$

Site A has the highest accident rate.

THE CORRECT ANSWER IS: (A)

530. Reference: Khisty, *Transportation Engineering*, p. 634.

Total vehicles entering intersection = 2,000 + 6,000 + 4,000 + 8,000 = 20,000 vpd.

$$\text{Rate} = \frac{(\text{number of accidents})(10^6)}{(\text{ADT})(\text{number of years})(365\text{ days/year})}$$

$$= \frac{(25)(10^6)}{(20{,}000)(9/12)(365)} = 4.6\text{ RMEV}$$

or

$$\text{Rate} = \frac{(\text{number of accidents})(10^6)}{(\text{ADT})(\text{number of days})}$$

Number of days during Jan. through Sept. $= 31 + 28 + 31 + 30 + 31 + 30 + 31 + 31 + 30 = 273$

$$\text{Rate} = \frac{(25)(10^6)}{(20{,}000)(273)} = 4.6\text{ RMEV}$$

THE CORRECT ANSWER IS: (A)

531. Reference: Chow, *Applied Hydrology*, p. 502.

Using Manning's equation:

$$Q = \left(\frac{1.49}{n}\right)(A)(R)^{2/3}(S)^{1/2}$$

$$A = \frac{\pi d^2}{4} \qquad R = \frac{d}{4} \text{ (for circular pipe)}$$

$$Q = \left(\frac{1.49}{n}\right)\left(\frac{\pi d^2}{4}\right)\left(\frac{d}{4}\right)^{2/3}(S)^{1/2}$$

$$43 = \left(\frac{1.49}{0.025}\right)\left(0.785\, d^2\right)\left(0.397\, d^{2/3}\right)\left(0.001^{1/2}\right)$$

$$d^{8/3} = 73.20$$

$$d = 5.0 \text{ ft}$$

THE CORRECT ANSWER IS: (B)

532. Reference: Chow, *Open-Channel Hydraulics*, p. 21.

Using Manning's equation:

$$Q = \left(\frac{1.49}{n}\right)(A)(R)^{2/3}(S)^{1/2}$$

$$Q = 1{,}800 \text{ cfs}$$

Given:

maximum depth = 4 ft

slope = 6%

$n = 0.012$

$$1{,}800 = \left(\frac{1.49}{0.012}\right)(A)\left(\frac{(A)}{WP}\right)^{2/3}(0.06)^{1/2}$$

$$(A)\left(\frac{(A)}{WP}\right)^{2/3} = 59.182$$

Trial and error at 4-ft depth:

Width	Area	Wetted Perimeter	$(A)\left(\frac{(A)}{WP}\right)^{2/3}$
6 ft	$4 \times 6 = 24$	$2 \times 4 + 6 = 14$	34.38
8 ft	$4 \times 8 = 32$	$2 \times 4 + 8 = 16$	50.80
10 ft	$4 \times 10 = 40$	$2 \times 4 + 10 = 18$	68.12

THE CORRECT ANSWER IS: (B)

532. (Continued)

For fewer calculations:

$A = w \times d$

$WP = 2d + w$

$$(w \times d)\left(\frac{w \times d}{2d + w}\right)^{2/3} = 59.182$$

$$(4w)\left(\frac{4w}{8 + w}\right)^{2/3} = 59.182$$

Use the same trial and error method as above by substituting w at 6, 8,10, etc. until greater than 59.182.

THE CORRECT ANSWER IS: (B)

533. Reference: Garber and Hoel, *Traffic and Highway Engineering*, 3rd ed., 2002, pp. 750–752.

Existing $t_c = 0.5$ hr (t_i)

10-yr storm

From chart, $I_1 = 3.25$ in/hr

Average $C = 0.36$ (given)

$Q_{exist} = C_1 I_1 A = (0.36)(3.25)(50) = 58.5$ cfs

Calculate C for proposed development:

$$= \frac{(10 \times 0.2) + (10 \times 0.35) + (18 \times 0.5) + (12 \times 0.65)}{50}$$

$$= \frac{2 + 3.5 + 9 + 7.8}{50}$$

$C_2 = 0.446$

Calculate new time of concentration:

$$\begin{aligned} t_2 &= t_1 \times 0.7 \\ &= 0.5 \times 0.7 \\ &= 0.35 \text{ hr} \end{aligned}$$

From chart for $t_2 = 0.35$ and 10-yr storm:

$I_2 = 4.1$ in/hr

$$\begin{aligned} Q_{proposed} &= C_2 I_2 A \\ &= (0.446)(4.1)(50) = 91.4 \end{aligned}$$

Increase $= 91.4 - 58.5 = 32.9$ cfs

THE CORRECT ANSWER IS: (B)

534. Reference: Garber and Hoel, *Traffic and Highway Engineering*, 3rd ed., 2002, pp. 843–846.

Since more than half of the sample is coarse grained (82% larger than No. 200 sieve) and more than half of the coarse fraction is a sand (72% passes the No. 4 sieve), the soil is classified as a sand. Furthermore, since the soil contains more than 12% fines (18% passing the No. 200 sieve), the soil is classified as a sand with fines. Therefore, the only appropriate choice among those given is "silty sand."

THE CORRECT ANSWER IS: (C)

535. Reference: Garber and Hoel, *Traffic and Highway Engineering*, 3rd ed., 2002, pp. 831–835.

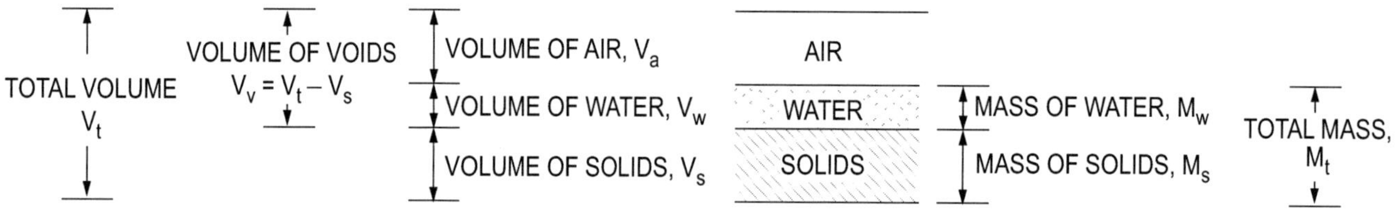

$$M_w = 338 - 228 = 110 \text{ g}$$

Density of water, $\rho_w = 1$ g/c.c.

Volume of water $\quad V_w = \dfrac{M_w}{\rho_w} = \dfrac{110 \text{ g}}{1.0 \text{ g/c.c.}} = 110$ c.c.

Volume of solids $\quad V_s = \dfrac{M_s}{SG\,(\rho_w)} = \dfrac{228 \text{ g}}{2.75(1 \text{ g/c.c.})} = 82.91$ c.c.

Volume of air $\quad V_a = 209 - 82.91 - 110 = 16.09$ c.c.

Void ratio $= e = \dfrac{V_w + V_a}{V_s} = \dfrac{110 + 16.09}{82.91} = 1.52$

Alternate solution:

$$e = \frac{V_v}{V_s}$$

$$V_v = V_t - V_s = 209 - V_s$$

$$V_s = \frac{M_s}{SG(\rho_w)} = \frac{228 \text{ g}}{(2.75)(1 \text{ g/c.c.})} = 82.91 \text{ c.c.}$$

$$V_v = 209 - 82.91 = 126.09 \text{ c.c.}$$

$$e = \frac{126.09}{82.91} = 1.52$$

THE CORRECT ANSWER IS: (A)

536. Reference: Chen and Liew, *The Civil Engineering Handbook*, 2nd ed., 1995, pp. 27-3, 27-4.

The standard penetration value (*N* value) is the number of blows required to penetrate the last 12 in.

N value (sample 7) = 5 + 6 = 11

THE CORRECT ANSWER IS: (C)

537. Reference: Garber and Hoel, *Traffic and Highway Engineering*, 3rd ed., 2002, pp. 807–808.

$$W = (\text{no. of lanes} \times \text{width}) + (2 \times \text{shoulder width})$$
$$= (2 \times 12) + (2 \times 6) = 36 \text{ ft}$$
$$N_c = 2 + 1 = 3$$
$$q_i = I_c\left[\frac{N_c}{W} + \frac{W_c}{WC_s}\right] + K_p$$
$$= 2.4\left[\frac{3}{36} + \frac{20}{36 \times 30}\right] + 0.05$$
$$= 0.29 \text{ ft}^3/\text{day/ft}^2$$

THE CORRECT ANSWER IS: (B)

538. Reference: Ricketts, Loftin, and Merritt, *Standard Handbook for Civil Engineers*, 5th ed., pp. 13-18–13-20.

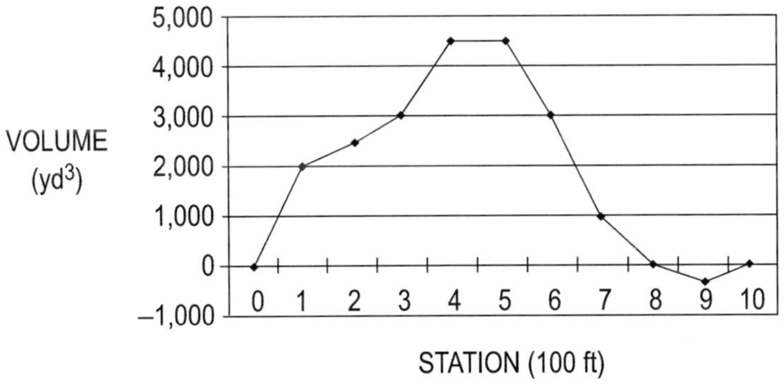

STATION	EXC.	EMB.	Δ	Σ
0+00	0	0	0	0
1+00	3,000	1,000	2,000	2,000
2+00	1,000	500	500	2,500
3+00	1,500	1,000	500	3,000
4+00	2,000	500	1,500	4,500
5+00	1,000	1,000	0	4,500
6+00	500	2,000	–1,500	3,000
7+00	0	2,000	–2,000	1,000
8+00	**0**	**1,000**	**–1,000**	**0**
9+00	500	1,000	–500	–500
10+00	500	0	500	0

THE CORRECT ANSWER IS: (C)

539. Reference: Ricketts, Loftin, and Merritt, *Standard Handbook for Civil Engineers*, 5th ed., pp. 13-18–13-20, or Nunnally, *Construction Methods and Management*, 5th ed., 2001, pp. 38-39.

Draw maximum ordinate for between balance line and curve in fill section at Station 6+00. Draw horizontal line through middle ordinate and measure distance of horizontal line.

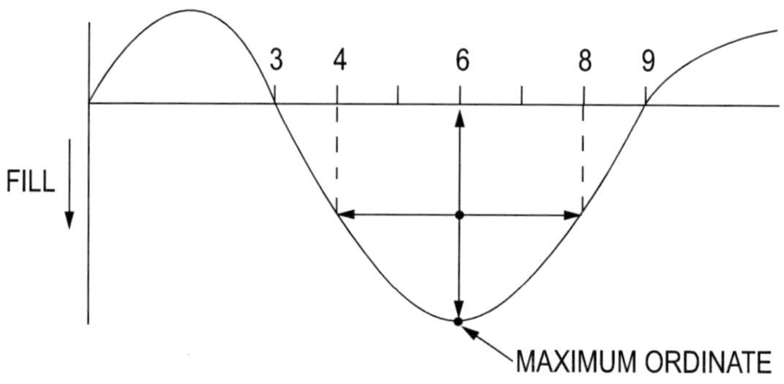

HAUL = 8+00 – 4+00 = 400 ft

THE CORRECT ANSWER IS: (B)

540. Reference: AASHTO, *Design of Pavement Structures*, 1993, p. II-14.

Calculate relative damage u_f:

$$\text{Wet } u_f = \left(1.18 \times 10^8\right)\left(3{,}000^{-2.32}\right) = 1.0115$$

$$\text{Dry } u_f = \left(1.18 \times 10^8\right)\left(10{,}000^{-2.32}\right) = 0.0619$$

$$\text{Calculate average } u_f = \frac{(1.0115 \times 4) + (0.0619 \times 8)}{12} = 0.378$$

Using chart in AASHTO (p. II-14) or the given u_f equation:

$$0.378 = 1.18 \times 10^8 \left(M_r^{-2.32}\right)$$

$$M_r = 4{,}582$$

THE CORRECT ANSWER IS: (C)

PE Sample Questions and Solutions Published by NCEES

Chemical
Civil: Construction
Civil: Geotechnical
Civil: Structural
Civil: Water Resources and Environmental
Electrical and Computer: Computer Engineering
Electrical and Computer: Electrical and Electronics
Electrical and Computer: Power
Environmental
Mechanical: HVAC and Refrigeration
Mechanical: Mechanical Systems and Materials
Mechanical: Thermal and Fluids Systems
Structural